THE NORTH AMERICAN PERCHING AND DABBLING DUCKS

The North American

Perching and Dabbling Ducks

Their Biology and Behavior

Paul A. Johnsgard

School of Biological Sciences
University of Nebraska–Lincoln

Zea Books, Lincoln, Nebraska: 2017

Abstract

This volume, the fourth in a series of books that collectively update and expand P. A. Johnsgard's 1975 *The Waterfowl of North America*, summarizes research findings on this economically and ecologically important group of waterfowl. The volume includes the mostly tropical perching duck tribe Cairinini, of which two species, the muscovy duck and the wood duck, are representatives. Both species are adapted for foraging on the water surface, mostly on plant materials, but typically perch in trees and nest in elevated tree cavities or other elevated recesses. This volume also includes the dabbling, or surface-feeding, duck tribe Anatini, a large assemblage of duck species that mainly forage on the water surface but nest on the ground, or only very rarely in elevated locations. Of this tribe, 12 species that regularly breed in North America are included, among them such familiar species as mallards, wigeons, pintails, and teal. Descriptive accounts of the distributions, populations, ecologies, social-sexual behaviors, and breeding biology of all these species are provided, together with distribution maps. Five additional Eurasian and West Indian species have been reported several times in North America; these have been included with more abbreviated accounts, but all 17 species are illustrated by drawings, photographs, or both. The text includes about 84,000 words and contains more than 1,000 references. There are also 12 distribution maps, 21 drawings, 28 photographic plates, and 58 anatomical or behavioral sketches.

Spot illustrations — page 1: Male and female wood duck; page 2: Northern pintails; page 10: Northern shoveler, male; page 12: Northern mallard, female and brood; page 22: Blue-winged teal, male in flight; page 23: Blue-winged teal, male swimming; page 120: Northern mallard, pair landing; page 187: North American green-winged teal, three-bird flight; page 227: Wood duck, male.

Text, photographs, and illustrations copyright © 2017 Paul A. Johnsgard

ISBN 978-1-60962-109-4

https://doi.org/10.13014/K27H1GGV

Composed in Adobe Garamond and Imprint MT Shadow types.

Zea Books are published by the University of Nebraska–Lincoln Libraries
Electronic (pdf) edition available online at http://digitalcommons.unl.edu/zeabook/
Print edition available from http://www.lulu.com/spotlight/unlib

UNL does not discriminate based upon any protected status.
Please go to unl.edu/nondiscrimination

Nebraska
UNIVERSITY OF
Lincoln

Contents

Maps

Figures

Photographs

Preface

Considering all of the nine major world-survey bird monographs that I have published, the earliest (1975) was my *Waterfowl of North America*. It was undertaken during an era of manual typewriters, carbon-copy manuscripts, and arduous literature searches that required making personal visits to sometimes distant cities and arranging for frustratingly slow interlibrary loans. I wrote my reference notes and copied the citations in longhand on 5-by-7-inch index cards, which were filed alphabetically or organized taxonomically in long cardboard boxes. Four decades later, my old manual and electric typewriters have now been long discarded, and I also gleefully trashed my endless boxes of yellowing index cards that I began assembling many decades ago. Those twentieth-century artifacts have since been happily replaced by computers and electronic searching programs via the Internet, both which have reduced my usual book-writing time from at least three years to less than one year.

Because of these technical advances, my idea of revising part of my *Waterfowl of North America* wasn't so daunting as when I first thought of it about three years ago. My initial idea was simply to write two books, updating the descriptive sections of the North American geese and swans (but later I added four other swan taxa beyond North America so as to include all of the world's swan taxa), and to assemble a new list of references that would supplement the approximately 400 in my original book. Both of these books were published electronically in 2016: the first, *Swans: Their Biology and Natural History*, with a text of nearly 50,000 words and slightly fewer than 700 citations, and the second, *The North American Geese: Their Biology and Behavior*, totaling about 60,000 words and having slightly more than 700 citations. Then I began work on a third volume, *The North American Sea Ducks: Their Biology and Behavior*, and it was completed by the autumn of 2016, with 92,000 words and about 900 references. The materials for this book were then placed in the highly capable hands of Paul Royster with Zea Books at the University of Nebraska–Lincoln Libraries, and it was published at the end of 2016.

This current volume, which describes North America's perching and surface-feeding ducks, was by then also underway and was essentially finished by the end of 2016. It comprises about 70,000 words, including more than 1,000 references. The text, photographs, maps, and drawings in all these volumes are mine, and are under my copyright. Although my 1960s behavior sketches are rather crude, I thought their visual informative values slightly outweighed their artistic limitations, and thus they too are also included. A final volume will include two breeding species of whistling ducks, five pochards, and two stiff-tailed ducks and describe all the remaining species of North American waterfowl.

    As for my many previous monographs that are already available in the DigitalCommons@University of Nebraska–Lincoln, I owe a huge debt of gratitude to Paul Royster, coordinator of Scholarly Communications in the Digital Initiatives and Special Collections department at the University of Nebraska–Lincoln Libraries, and to his editorial staff for seeing this book through to completion. I additionally am much indebted to his invaluable editor, Linnea Fredrickson. Thanks also to the university's librarians for their cheerful help in locating and providing me with endless reference materials.

                                        Paul A. Johnsgard
                                        Foundation Regents Professor Emeritus
                                        Biological Sciences
                                        University of Nebraska–Lincoln

Introduction

If almost anybody were asked to name a duck species with which they are familiar, that person is almost certain to mention the mallard. Mallards are probably the most abundant wild duck species in the world, and also are among the species most widespread, the most widely prized by sport hunters, and the most widely domesticated of all the approximately150 waterfowl species.

Mallards are but one of some 20 species of worldwide duck species that are mostly temperate-zone in climatic distribution, and that mostly "dabble" in shallow water for their foods, which are predominantly obtained from vegetable sources. Also called surface-feeding ducks, they are close relatives of a smaller group of similar mostly surface-feeding ducks known as perching ducks that are more tropically oriented, usually are found in woodland habitats, and frequently nest in elevated cavities. Of these species, the North American wood duck and its close relative the Asian mandarin duck (*Aix galericulata*) are the best-known examples, and are also often regarded as the most beautiful of all the world's species of ducks, geese, and swans.

Because of their diversity, abundance, and high economic values, the popular and technical research literature of these waterfowl is enormous. In preparing this book, which is mostly an updating of my 1975 volume *The Waterfowl of North America*, I have surveyed a large number of publications, trying to bring in newer research findings of the ecology, behavior, and populations of these birds. My earlier book had slightly over 100 literature citations dealing largely or entirely with dabbling and perching ducks, whereas this present volume has approximately 900. Luckily, the majority of sources can now be readily found through Internet searches. Because of the length of the bibliography, I have subdivided it in a largely taxonomic and geographic manner, so that, when trying to find the title of a publication mentioned in the text, it might be necessary to search in two or more potential locations. For papers that focus on only two primary species, I have included the citation under both species' categories, but if three or more species were the focus of a study, it should be searched for under multispecies studies or another of the broad-category topics.

My sketches of dabbling duck displays were based on my studies of 16 mm cine frames and have been reprinted from my 1965 *Handbook of Waterfowl Behavior*. That book provides written descriptions of all the displays I have illustrated and can be freely accessed at http://digitalcommons.unl.edu/bioscihandwaterfowl/7/.

Two major reference books on waterfowl have been published since 1975, including a two-volume world survey of waterfowl, *Ducks, Geese and Swans*, which was the marvelous swan-song effort by a long-time friend, Dr. Janet Kear (1933–2004), and which was published in 2005 shortly after her untimely death. Additionally, Professor Guy Baldassarre (1953–2012) provided a splendid major revision of F. H. Kortright's classic 1942 *Ducks, Geese, and Swans of North America*, which was likewise sadly published posthumously (2014).

Equally important individual species monographs by various authors of all the North American ducks have also appeared since the 1990s, through the joint multiyear *The Birds of North America* project of the American Ornithologists' Union and the Academy of Natural Sciences of Drexel University. All of these, including some updated versions, have recently become available online through the cooperation of Cornell University's Laboratory of Ornithology (see Birds of North America online at http://bna.birds.cornell. edu/bna/species/).

I. Introduction to the Perching and Dabbling Ducks

Perching ducks and dabbling ducks are two of the major subdivisions of the waterfowl subfamily Anatinae (Johnsgard, 1979a). All these "anatine" species also have a distinctive structural feature in that they exhibit a row of scales that are vertically aligned (scutellated) just above the base of the middle toe (Fig. 1), rather than having the web-like (reticulated) scale pattern found in swans, geese, and whistling ducks.

Unlike the whistling ducks, swans, and true geese of the subfamily Anserinae (the "anserines"), in all of the approximately120 species of typical ducks the sexes are highly diverse in their adult breeding plumages, vocalizations, and sexual behavior patterns. These sex-based differences can be attributed to the weaker, less permanent pair-bonds characteristic of anatine ducks, with a resultant need for renewing pair-bonds each year, and the consequent advantages of having sufficient intersexual and interspecies differences in behavior, vocalizations, and plumage markings so as to achieve appropriate mate choice.

In common with all other waterfowl (and most other flying birds) all anatine ducks have 10 outer flight feathers (primaries) per wing, which are attached to the fused hand bones (Fig. 1). The inner flight feathers that are attached to the arm bones (ulnas) are termed secondaries and typically number about 15. Of these, the innermost 5 to 7 secondary feathers (often called tertials) are much more variable in shape and size, and imperceptibly merge inwardly with the back feathers. In swans and geese ("anserine" species), the secondary feathers are more uniform in shape but are more variable in number, often exceeding 20. Nearly all anatine species also have 16 to 18 tail feathers (rectrices), whereas in geese and swans 16 to 24 rectrices are present.

Adult males of the large anatine assemblage also have a unique tracheal (windpipe) structure, the tracheal bulla. The bulla is a variably inflated bony chamber located where the bronchial tubes originating at the lungs connect with the trachea. Here two pairs of vibratory membranes also occur, supported by the syrinx, a bony or cartilaginous structure that is unique to birds. The syrinx generates sounds (vocalizations) that are controlled by air pressures originating from the lungs and varied muscular tensions on the syringeal membranes that result in variable membrane vibration characteristics. Acoustic differences in acoustic loudness (amplitude), pitch (sound frequency), and harmonics (overtones) are thereby generated and modulated. The tracheal bulla somehow further modifies the vocal characteristics of these sounds, and interspecies variations in the size and structure of each species' trachea and tracheal bulla might facilitate greater interspecies acoustic diversity.

The species of this subfamily are currently grouped into a still-contested number of smaller "tribes," most of which include one or more native North American species. On the basis of a cladistic analysis of largely morphological traits, rather than on biological characteristics of living animals, Livizey (1997) merged the perching and dabbling duck groups. However, my behavioral research strongly supports a separation of the

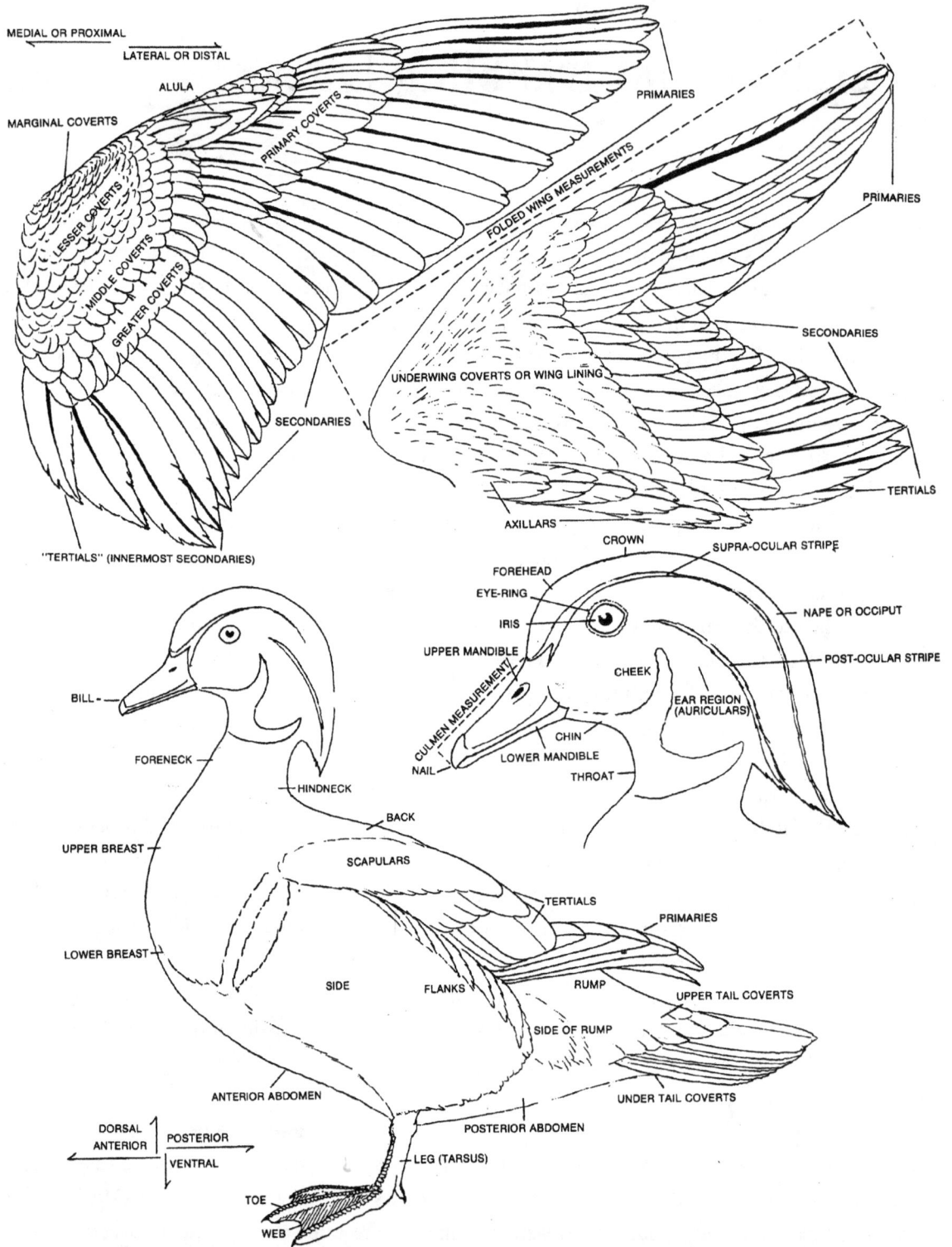

Fig. 1. External topography of a perching duck (wood duck).

dabbling and perching ducks that closely conforms to Jean Delacour and Ernst Mayr's 1945 taxonomy. The perching ducks are thus here recognized as a distinct tribe that is very closely related to the dabbling ducks but is genetically different enough as to prevent the production of fertile hybrids, or sometimes even hybridization itself.

Perching Ducks

The perching ducks and a few related goose-like forms that make up the tribe Cairinini are a diverse array of some 14 species that are largely subtropical to tropical in distribution. Although they vary in size from as little as about half a pound in the pygmy geese (*Nettapus*) to more than 20 pounds in the African spur-winged geese (*Plectropterus*), all the species possess some associative features. These include notable perching abilities, preferential cavity nesting in trees, relatively sharp claws, and long, broad tails that presumably increase braking effectiveness when landing and perching on branches. They also have relatively long incubation and fledging periods, and prolonged breeding seasons. Most perching ducks also exhibit extensive iridescent coloration on their body plumage, especially on their inner wing feathers. This conspicuous coloration is often exhibited by females as well as males and is usually concentrated in a group of iridescent or otherwise conspicuous secondary wing feathers called the speculum (from Latin, a mirror). As a result, this tribe of waterfowl includes some of the most beautifully arrayed species of the entire waterfowl family.

The temperate-zone North American wood duck and the closely related Asian mandarin (*Aix galericulata*) are examples of the most highly ornamental perching ducks. The pair-forming behaviors of these ducks are notably complex, and their seasonally elaborate plumage patterns are perhaps a result of the frequent intense competition among males for mates during the brief temperate-zone pair-bonding period. As a result, it is easy to identify the distinctively plumaged males of most perching ducks, whereas the females of these closely related species are extremely similar in appearance.

In all perching ducks the males assume the initiative in pair-forming activities. Males of most have seasonally brief pair-bonds, are usually much more colorful than the females, and have more sexually divergent appearances and behavior patterns than occur among species with more permanent pairing. Females of most perching ducks exhibit more subdued and concealing plumage patterns, in association with their egg-laying and incubation responsibilities.

Some large-bodied tropical perching ducks have notably weak pair-bonds, such as the muscovy duck, spur-winged goose (*Plectropterus gambensis*), and comb duck (*Sarkidiornis* spp.). These species exhibit greater relative male strength and size than females; adult male muscovy ducks average about 2.5 times as heavy as females. Among these species, intense male-to-male aggressiveness seems to be more important than elaborate plumages and complex male courtship displays in determining an individual male's sexual success, but males play no further role in influencing breeding success.

In contrast, long-term pair-bonding is typical of several tropical, small-bodied perching ducks such as the ringed teal (*Callonetta leucophrys*), Amazon teal (*Amazonetta brasiliensis*), and pygmy geese (*Nettapus* spp.). In males of these species there are no seasonal variations in their generally less iridescent plumages, male aggressive and pair-forming behaviors are infrequent and inconspicuous, and male participation in brood care is probably typical of all.

Dabbling Ducks

The dabbling, or surface-feeding, ducks are a group of about 36 species of mostly freshwater ducks that occur throughout the world. Many of them are cold-temperate or Arctic-breeding species that are found on freshwater ponds, lagoons, marshes, slow-moving rivers, or other shallow wetlands. Like perching ducks, and in contrast to diving ducks (the pochards, sea ducks, and stiff-tailed ducks), they are fairly mobile on land, and often forage on land. However, most foraging is done on the water, either on or just below the water surface.

Diverse foraging techniques are used, such as "dabbling" for food items while swimming or standing at the water's edge, by sieving very small particles from along the water surface by straining them out with their bill's comb-like lamellae, or by submerging the head and "tipping-up" (or "up-ending") to reach deeper sources of food. In most species the foods consumed are predominantly of vegetable matter, whereas most diving ducks, especially sea ducks, primarily eat materials of animal origin. However, most dabbling and perching ducks regularly dive for food in shallow water, and all can readily dive to escape danger. Both perching and dabbling ducks can also take flight abruptly from small water areas or land without first running fast enough to attain flight speed. However, dabbling ducks have reduced capabilities for landing in trees, and also have limited perching capabilities as compared with perching ducks, which typically have longer, broader tails that assist in aerial braking, and longer claws on their toes.

Also unlike perching ducks, whose iridescent body plumage is often extensive in both sexes, such feather specialization among dabbling ducks is mostly limited to a conspicuous speculum pattern on the secondary wing feathers of most species, plus variable amounts of head iridescence on the males of some. Colorful male plumages in dabbling ducks are probably "expensive" to maintain in terms of their visibility and increased probability of detection by predators. Thus, except when the birds are in flight the otherwise conspicuous wing speculum is effectively hidden by other overlaying feathers. Furthermore, among insular or historically isolated populations having reduced dangers of interspecies hybridization, as is true of the American black, mottled, and Florida ducks, the males have evolved permanent concealingly patterned plumages closely resembling those of females.

The dabbling ducks are the most abundant and familiar of all North American ducks, and include such popular sporting species as mallards, pintails, wigeons, and various teals. They range in size from less than a pound to more than three pounds and are among the most agile of waterfowl in flight, relying on maneuverability rather than high speed to elude danger. In most respects the surface-feeding ducks closely conform with perching ducks in their basic anatomy and biology but differ from them in that they are nearly all ground-nesters and are poorly adapted for perching or nesting in elevated cavities.

Like some tropical perching ducks, a long-term pair-bonding pattern also occurs in several South American species of dabbling ducks, where breeding might occur year-round, or opportunistically whenever environmental conditions permit. Such long-term bonding and biparental brood care traits are present at least in the crested duck (*Lophonetta specularioides*), bronze-winged duck (*Anas specularis*), and Chiloe wigeon (*Anas sibilatrix*). In some other subtropical South American *Anas* species such as the white-cheeked pintail (*Anas bahamensis*), yellow-billed teal (*Anas flavirostris*) and brown pintail (*Anas georgica)* the male often accompanies the female during the brood-rearing period, and possibly defends her, but does not actively assist in brood care and protection.

For example, under the influence of a sedentary population and extended and variable breeding seasons the mating system of the white-cheeked pintail was found by Sorenson (1992) to be highly variable. Most males were monogamous, but some males that were unusually effective in guarding their mates were polygynous ("male quality polygyny"). Some birds formed monogamous, long-term pair-bonds lasting two years or more, while other pairs "divorced" after their first breeding season, Only females provided brood care, but some males escorted and guarded their mates into at least part of the brood-rearing season.

Molts and Plumages of Perching and Dabbling Ducks

Following the initiation of incubation, most male perching and dabbling ducks abandon their incubating mates and begin their postnuptial (or "prebasic") molt. They then become temporarily flightless while replacing their flight feathers, and most species also acquire a more inconspicuous ("basic") plumage. Thus, unlike swans and geese, most ducks have two plumages, and two intervening body feather molts, per year. This double molt is more apparent in males having sexually dimorphic plumages, since the resulting basic plumage (usually called the "eclipse" plumage) is both less colorful and female-like than is the breeding (alternate) plumage.

Although in all the duck species that have so far been studied the female also has a comparable summer molt and corresponding nonbreeding plumage, in most cases this plumage is so similar to the breeding plumage that separate descriptions are not necessary. In most cases the basic plumage of males is held for only a few months, allowing the male to regain the more distinctive alternate plumage associated with pair formation as early as possible, usually during autumn or early winter.

In some cases (e.g., ruddy duck, Baikal teal, blue-winged teal), however, the male's breeding plumage is not regained until well into spring so that "summer" and "winter" plumages may be more seasonally descriptive. Any formal plumage-naming protocol is further complicated in the long-tailed duck, which has a third partial molt in the fall (most apparent in males) and which is restricted to the scapular region. Except in such special cases, the two major plumages of the male follow long tradition and are referred to in the species accounts as "nuptial" and "eclipse" plumages, while the "adult" plumage of females here refers to both of the comparable breeding and nonbreeding plumages as well as to any minor female molts and plumages.

North American Species Diversity, Rarities, and Hybridization

Although a great deal of diversity in bill shapes and male/female plumage pattern differences exists among the dabbling and perching ducks (Figs. 2 and 3), most biologists now agree that recognition of a single genus (*Anas*) is most representative of the close relationships among the North American dabbling ducks. Separate generic recognition for the shovelers (*Spatula*), wigeons (*Mareca*), pintails (*Dafila*), gadwall (*Chaulelasmus*), teal (*Querquedula*), and others obscures the obviously close relationships that exist within this group and helps explain the reports of their occasional hybridization in the wild (Johnsgard, 1960). With few exceptions, I have followed the taxonomic sequence and nomenclature used in my 1979 classification of the Anatidae, as well as in various earlier publications (Johnsgard, 1961a, 1961b, 1965, 1979a).

The total number of North American breeding species of anatine ducks is somewhat uncertain but is at least nine. The Mexican population of the northern mallard has at times been considered a separate species from the common mallard, as currently are two other mallard-derived and once geographically isolated populations: the Florida duck and mottled duck. Genetically and behaviorally, all of these "southern mallards" (Johnsgard, 1975) are barely distinct from northern mallards. Thus it is a matter of tradition rather than of taxonomic accuracy to retain two of them as specifically distinct from the northern mallard, and only one as a mallard subspecies, as I have done here.

Additionally, there are several thousand sighting reports of the Eurasian wigeon in continental North America, and the occurrence of occasional wild hybrids involving the American wigeon, suggesting that it very probably breeds occasionally in continental North America (Fournier and Hines, 1996). More rarely, periodic Pacific coast sightings of the falcated duck proves that it too belongs on the list of North American birds, and in recent years there have also been multiple sighting records for the Baikal teal, white-cheeked (Bahama) pintail, and garganey. It seems very likely that at least some of these records have involved wild birds rather than escapes from captivity; thus, abbreviated species accounts for these five species are included here.

In addition to these rare "foreign visitors," waterfowl hybrids are surprisingly common among North American dabbling ducks and pose identification and species-counting problems. The now almost rampant hybridization between the abundant northern mallard and the relatively small populations represented by the Mexican, mottled, and Florida ducks is gradually encroaching on their genetic identities. The intensity of hybridization between the northern mallard and the American black duck is only slightly less, in terms of the increasingly evident genetic threats that the mallard poses for the American black duck.

Within the genus *Anas* there appears to be a high rate of hybrid fertility, with more than 30 interspecific crosses known to be at least sometimes fertile (Johnsgard, 1960). However, such fertility appears to be absent among the numerous crosses that have been reported between *Anas* species and the wood duck. Some earlier comments have been made on the possible significance of the surprisingly numerous known hybrids involving the wood duck and various *Anas* species (Dilger and Johnsgard, 1959), although all these perching–dabbling duck hybrids have proven sterile.

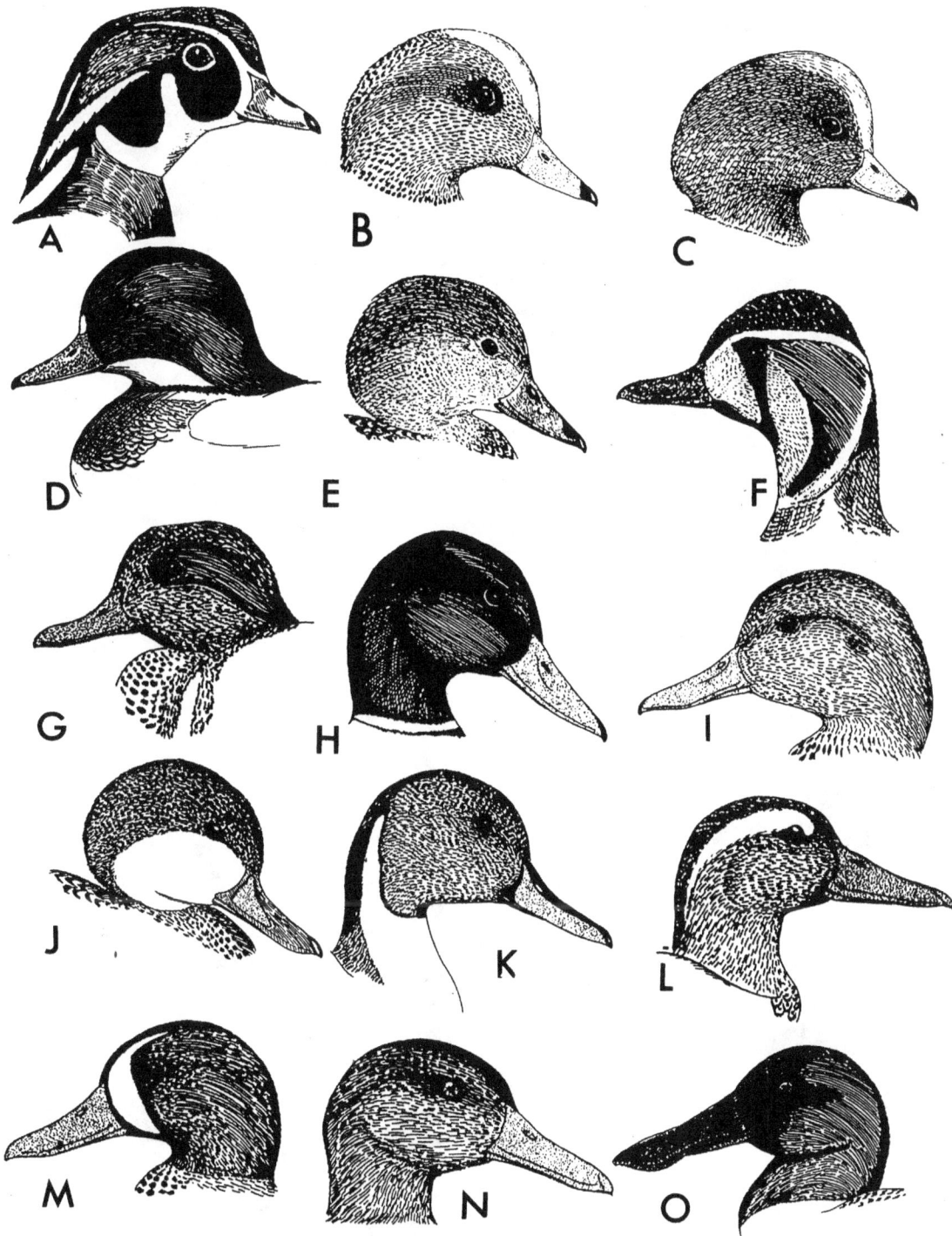

Fig. 2. Head profiles of male perching and dabbling ducks, including (A) wood duck, (B) American wigeon, (C) Eurasian wigeon, (D) falcated duck, (E) gadwall, (F) Baikal teal, (G) green-winged teal, (H) northern mallard, (I) American black duck, (J) white-cheeked pintail, (K) northern pintail, (L) garganey, (M) blue-winged teal, (N) cinnamon teal, and (O) northern shoveler.

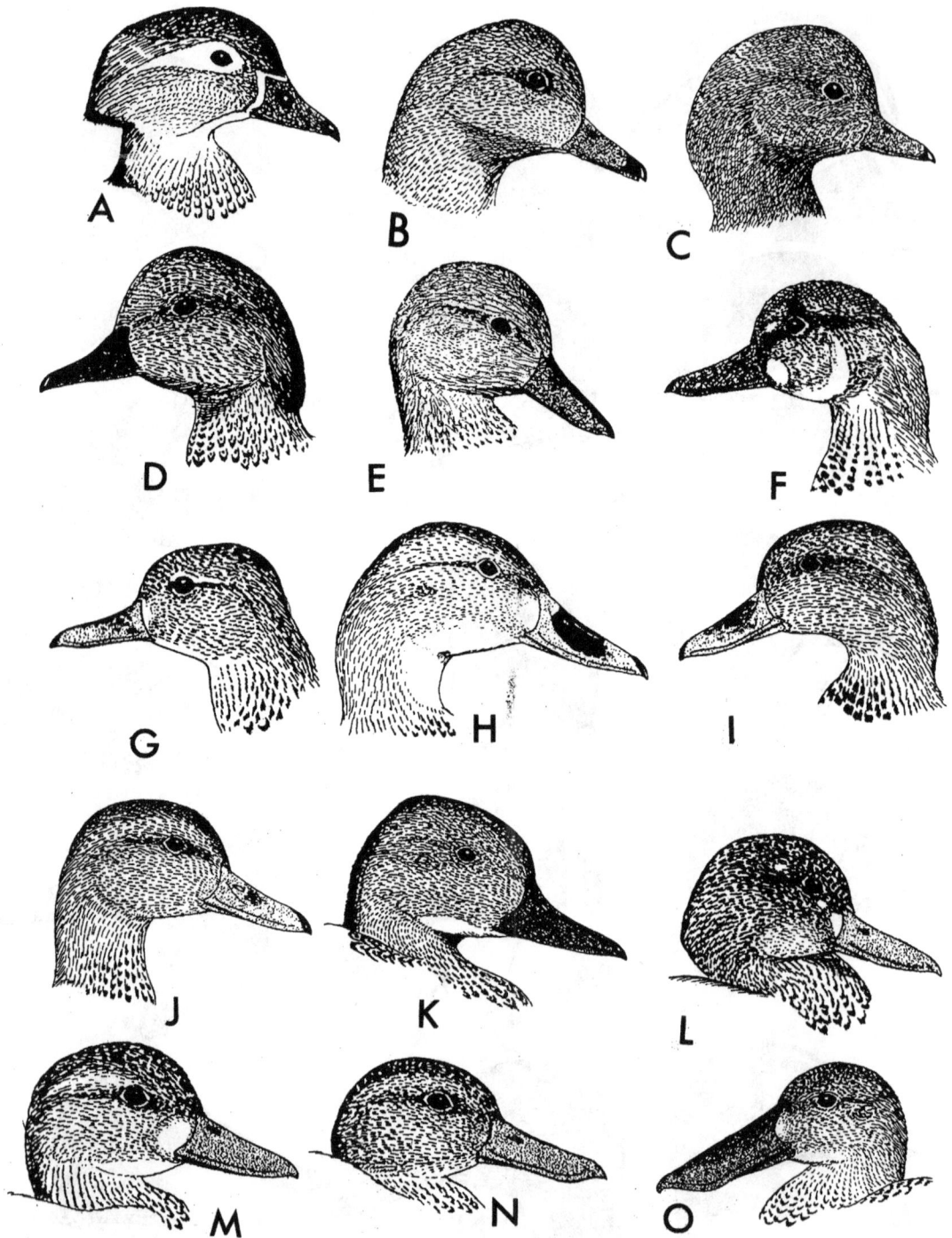

Fig. 3. Head profiles of female perching and dabbling ducks, including (A) wood duck, (B) American wigeon, (C) Eurasian wigeon, (D) falcated duck, (E) gadwall, (F) Baikal teal, (G) green-winged teal, (H) northern mallard, (I) American black duck, (J) Florida duck, (K) northern pintail, (L) garganey, (M) blue-winged teal, (N) cinnamon teal, and (O) northern shoveler.

Not all of the approximately three dozen species of *Anas* have yet been studied ethologically as to their comparative pair-forming behavior, but enough have been observed to provide a general understanding of the variety and functions of the many evolved male sexual displays. Pair-forming displays are behavioral patterns that have become "ritualized" over time through natural selection and have thus acquired important social communication functions in facilitating pair formation (mate-choice behavior) and pair-bonding (mate-retention behavior) activities. I described the social behavior of waterfowl in a previous monograph (Johnsgard, 1965), and space here does not permit a repetition of detailed display descriptions. Yet, the sexual displays of perching and dabbling ducks are both highly important biologically and ethologically, so the more interesting ones will be mentioned, and a few of them illustrated, in the species accounts that follow.

II. Species Accounts

Tribe Cairinini (Perching Ducks)

Muscovy Duck
Cairina moschata (Linnaeus) 1758

Other vernacular names. Musk duck, pato real

Range. From the lower Rio Grande valley of Texas south through tropical lowlands of Mexico, the tropical woodlands of Central and South America to Peru on the Pacific slope, and Argentina on the Atlantic slope. The species is nonmigratory and relatively sedentary.

Subspecies. None recognized.

Measurements. *Folded wing:* Delacour (1954): Males 300–400 mm, females 300–315 mm. Palmer (1976): Males 345–408 mm (average of 9, 385 mm); females 295–318 mm (average of 4, 307 mm).
   Culmen (bill): Delacour (1954): Males 65–75 mm, females 50–53 mm. Palmer (1976): Males 60.9–76.2 mm (average of 9, 67.9 mm); females 47.2–54.3 mm (average of 4, 51.4 mm).

Weights (mass). Leopold (1959): Males 4.39–8.82 lb. (1,990–4,000 g), females 2.43–3.24 lb. (1,100–1,470 g). Gómez-Dellmeier and Crugan (1989): 8 males, average 3,077 g; 12 females, average 1,689 g.

Identification

In the hand. Any large, predominantly blackish duck with a rather broad, truncated tail measuring more than 100 mm and bare skin on the face is of this species. Domesticated varieties, which are sometimes shot by hunters, may vary greatly in coloration but usually are quite large and obviously of domestic origin.

In the field. Within its Central and South American range, the muscovy is largely confined to coastal rivers and shallow lagoons, often in or near tropical forests. Although sometimes foraging in open situations, muscovies usually return to timbered areas to rest and roost. The blackish body coloration is evident when it is either standing on land or swimming in water, but the white upper wing-coverts are usually not apparent. In flight, the white under wing-coverts and the variably white pattern of the upper wing surface contrast strongly with the otherwise dark body. In spite of their size, muscovies fly swiftly and strongly, often producing considerable wing noise. Otherwise, they are quite silent, both in flight and at rest.

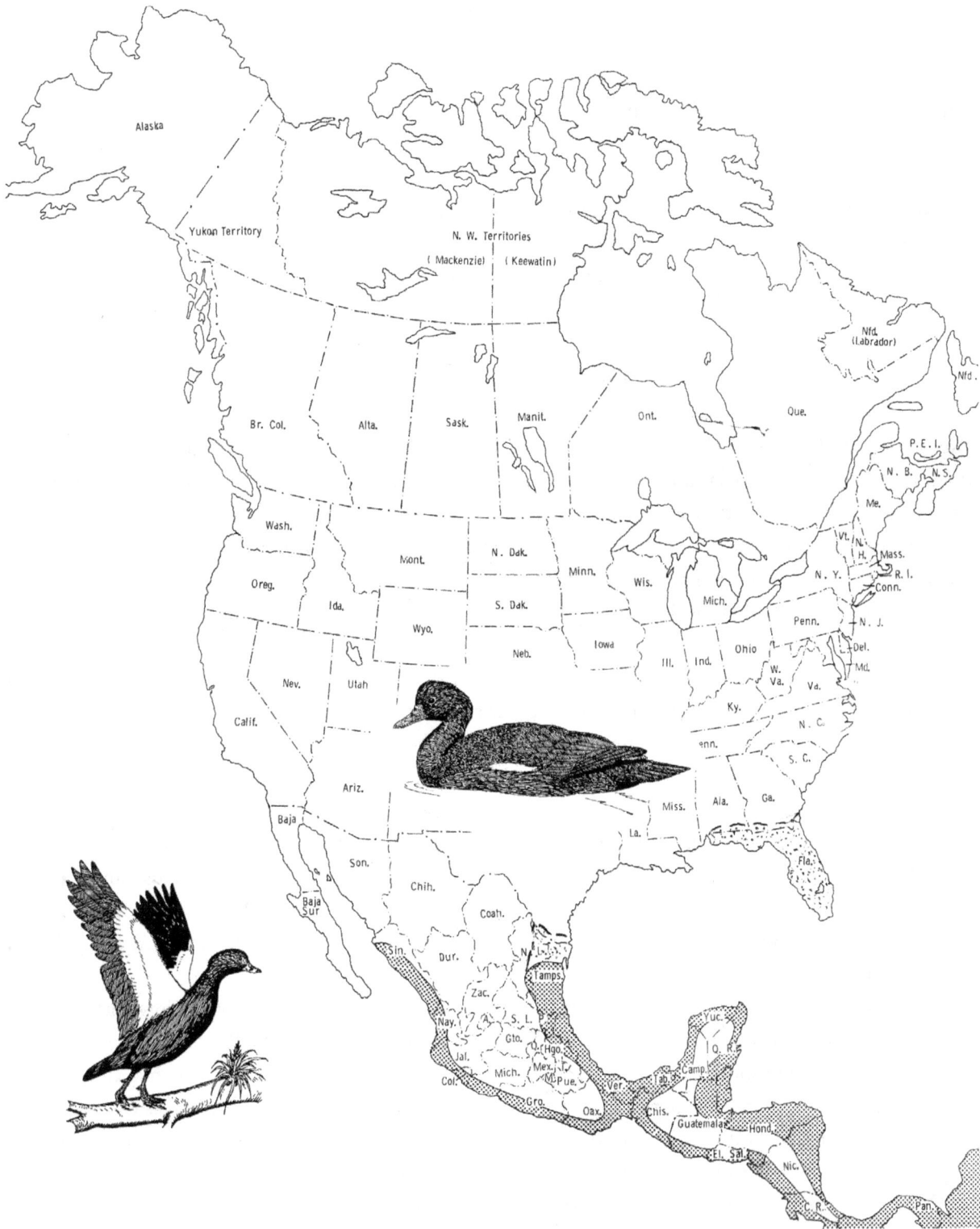

The residential (shaded) and acquired (stippled) range of the muscovy duck in North and Central America.

Age and Sex Criteria

Sex determination. In adults, the strong size dimorphism and caruncles on the head and bill of the male make sex determination simple. A culmen length in excess of 55 mm and the presence of naked skin on the face are indicative of a male.

Age determination. First-year males are less glossy overall, the lores feathered. The amount of white present on the upper wing surface and the size of the caruncles on the male's bill increase with age until at least the second year. Sexual maturity is attained in the first year among captive birds, but the situation in wild muscovies is not certain.

Distribution and Habitat

Breeding distribution and habitat. The natural North American breeding distribution of the muscovy duck is limited to the lowland portions of Mexico, from central Sinaloa on the west and Nuevo Leon on the east southward and eastward along both coasts with the exception of those portions of the Yucatan Peninsula that lack suitable rivers and lagoons (Leopold, 1959). Until the 1990s there were no records of the species' natural occurrence in the United States, but attempts have been made as early as the 1960s to establish this species in Florida, using offspring of wild stock from South America. Feral populations have since developed in Florida, from both wild and domestic sources. By 2010 muscovies had developed feral populations around many Florida cities, and during the 2008–2009 Audubon Christmas Bird Count the greatest number seen at any single location was 440, at Naples, Florida. In recent years muscovies have been reported from Florida during all recent Audubon Christmas counts, and they have reported from virtually all of Florida's counties. They also are increasingly being reported from other Gulf coast states.

Wild-type muscovies were first seen in Texas in 1998, and by 2010 they were regularly seen along the lower Rio Grande in southern Texas, especially around Santa Ana National Wildlife Refuge and in the vicinity of Falcon Dam, Zapata County. The first known Texas nesting of wild-type birds occurred in 2004 in Hidalgo County. Some population dispersion occurred in 2010 following Hurricane Alex (Lockwood and Freeman, 2014).

The muscovy's residential range also extends southward through virtually all of the lowland regions of Mexico, from Nuevo Leon and Tamaulipas on the Gulf coast and Sinaloa on the Pacific coast south to Guatemala and Belize, and south through both coastal lowlands of Central America to Panama. It also ranges over much of tropical South America from Colombia south, especially the forested areas east of the Andes Mountains, but extends south to northern Peru on the Pacific slope. Its southern limits on the Atlantic slope are reached near Tucuman, Santiago del Estero, and Santa Fe, Argentina.

The species' breeding habitat consists of rivers, lagoons, marshes, and similar areas of water at relatively low altitudes that are associated with forests or heavy woodland. Slowly flowing rivers associated with tropical

Fig. 4. Male and female muscovy ducks.

forests as well as backwater swamps associated with such rivers seem to represent their preferred habitat. Trees are used for nocturnal roosting.

Population. Very little information is yet available on the biology and regional populations of wild birds, but a world population estimate of 100,000 to 1,000,000 muscovies has been suggested (Kear, 2005). Given the rate of tropical forest destruction in Central and South America, the lower number seems more realistic.

Wintering distribution and habitat. There are no indications of migratory movements in this species, which occurs in climates affected little, if at all, by seasonal temperature fluctuations. During dry seasons the birds often move into rainforests, coastal swamps, or lagoons.

General Biology

Age at maturity. Age at maturity is not yet established for wild birds, but domesticated muscovy ducks regularly breed in their first year of life, so the same is likely for wild birds.

Pair-bond pattern. Current evidence indicates that muscovies virtually lack pair-bonds, the matings occurring promiscuously, and, except during the limited period of female receptiveness, there is little close association between the sexes. This is apparently true of captive birds (Johnsgard, 1965), and a few observations on wild birds indicate that such a polygynous social pattern exists (Delacour, 1959). However, observations on wild birds in Mexico (Rojas, 1954), and on some captive birds (Sibley, 1967) have stated that muscovies are regularly seen in pairs during the breeding season.

Nest location. Nests are usually located from 3 to 20 meters high, in tree hollows or among palm leaves. However, nests located among rushes at ground level have been reported in Argentina, and nests are sometimes made among palm leaves (Phillips, 1922). In most cases little or almost no down is present in the nest. Nest boxes are also readily accepted as nest sites if the entrance hole is about 21 centimeters in diameter (Woodward and Bolen, 1984).

Clutch size. The normal clutch size is probably 8 to 9 eggs, but apparent dump-nesting sometimes results in clutches twice this size, or even larger (Phillips, 1921). Leopold (1959) noted a range of 8 to 14 eggs, and Markum and Baldassarre (1989) reported that nine clutches in nest boxes ranged from 9 to 15 eggs, averaging 12.6 eggs. Like many hole-nesting species, dump-nesting (or parasitic egg-laying) is apparently common in the muscovy duck. Markum and Baldassarre (1989) reported that four such multiple clutches averaged 17.7 eggs, and ranged from 15 to 21 eggs. Two of these clutches also contained some eggs from black-bellied whistling ducks.

Incubation period. A 35-day period has been reported for eggs from captive birds (Delacour, 1959; Lack, 1968). Markum and Baldassarre (1989) reported an incubation period of 30 to 31 days for seven successful nests by wild females using nest boxes.

Fledging period. Baicich and Harrison (1997) reported a fledging period of 70 days. Hoffmann (2005) estimated it to be about 3.5 months, or slightly over 100 days. This latter fledging duration would seem improbably long, but it is also typical of some of the larger tropically oriented perching ducks, such as the closely related white-winged wood duck (*Cairina scutulata*) (Kear, 2005).

Nest and egg losses. Markum and Baldassarre (1989) reported a 77 percent hatching success among 13 nests in nest boxes, and a hatching success of 73 percent among seven normal-sized clutches. A 59 percent hatching success resulted from eggs that had been deposited in 3 nest boxes by multiple females. Woodyard and Bolen (1984) reported a 75 percent nesting success among four nests.

Muscovy duck, adult male

Juvenile and adult mortality. Almost no information is available about muscovy mortality. Young birds are reportedly sometimes taken by crocodilians (Donkin, 1989). After their first year, it seems possible that at least males might have a rather low natural mortality rate because of their unusual size and strength.

General Ecology

Food and foraging. Phillips (1922) summarized the limited information on food available at the time of his research. Items reported taken included small fish, insects, small reptiles, and water plants. Termites are said to be a favorite food, and their nests are sometimes torn open by the birds in search of them. Muscovies have also been observed chasing small crabs and feeding on water-lily seeds and the roots of *Mandioca*. Woodyard and Bolen (1984) reported that the foods of 15 adults from Veracruz were mainly water lily (*Nymphaea*) and mangrove (*Avicennia nitida*) seeds. About 30 percent of their total food volume was from animal sources, mostly insect larvae and other invertebrates. Wetmore (1965) noted that the stomachs of two birds from Panama contained various seeds, including those of pickerelweeds (Pondeteriaceae) and sedges (*Fimbristylis*).

I have never observed captive muscovy ducks diving; they spent much of their time foraging on land, presumably for seeds and insects. Although fish have been reported as part of their diet, it seems unlikely that they would be able to capture them under normal conditions, since muscovies are bulky and rather awkward birds.

Sociality, densities, territoriality. During the breeding season, males are highly aggressive toward one another, and such behavior no doubt tends to disperse the breeding population. A single male is often associated with more than one female, and perhaps such females might sometimes nest in close proximity. In spite of strong male-to-male hostility, there is no evidence of typical territorial behavior (defense of an area), and no estimates of breeding densities are yet available.

Interspecific relationships. Not enough is known of the ecology of this species to speculate on its possible competitors and enemies, which might include large hawks and owls, crocodilians, fish, and perhaps predatory mammals. The comb duck (*Sarkidiornis melanotos*) is a fairly closely related tropical forest species that also nests in cavities, but the ecological relationships between these two ducks are still obscure. Comb ducks seemingly occupy more open country than do muscovy ducks and are thought to be less dependent on undisturbed forests.

General activity patterns and movements. Outside the breeding season, muscovies usually gather in groups ranging from a few to 50 or more birds, wandering about rather extensively (Monroe, 1968). The birds typically fly during morning and evening hours (Wetmore, 1965), often spending the warmer parts of the day resting along the shore. At night they typically retire to tree roosts, with as many as a dozen or more birds sometimes roosting in a single tree (Phillips, 1923).

Social and Sexual Behavior

Flocking behavior. Most observers have reported that wild muscovies are usually found in small groups of about a half dozen birds but occasionally are in larger groups. These groups are not closely coordinated and on disturbance will often disperse in all directions. Perhaps the advantages of common roosting behavior tend to maintain flocking behavior outside the breeding season; at least pair-bonds and family bonds do not seem to be sufficiently strong as to facilitate such flocking behavior. The notable aggression among males, at least during the breeding season, probably tends to keep group size low.

Pair-forming behavior. No definite pair-bonds have been found among captive or domestic muscovy ducks. The rather simple display of the male serves both as aggressive signals toward males and as sexually oriented signals toward females. At such times he utters a soft breathing or hissing note, simultaneously raising his crest, moving his head slowly forward and backward, shaking his tail, and holding his wings slightly away from the body. Females normally respond to this display by fleeing, sometimes uttering a simple quacking note. I have never observed any female behavior that could be interpreted as inciting behavior, and no other type of apparent pair-forming behavior has been observed by me (Johnsgard, 1965).

Muscovy ducks, adult pair

Copulatory behavior. According to most observers, copulation in this species normally takes the form of apparent rape, with the male chasing and eventually overpowering the much smaller female. However, during the egg-laying period the female may actively solicit copulation, assuming a prone posture on the water and waiting thus as the male performs his sometimes rather lengthy precopulatory behavior, which consists of characteristic head movements and pecking the female's back feathers. After treading, the female bathes, but no definite male postcopulatory displays have been described (Johnsgard, 1965).

Nesting and brooding behavior. Not yet studied in detail, the muscovy's nesting and brooding behavior is apparently rather similar to that of the wood duck and other hole-nesting perching ducks, such as comb ducks (*Sarkidiornis*) (Markum and Baldassarre, 1989).

Postbreeding behavior. Other than the fact that considerable wandering by wild birds occurs during the nonbreeding season, almost nothing is known of the postbreeding stage in the life cycle of wild muscovy ducks.

Wood Duck
Aix sponsa (Linnaeus) 1758

Other vernacular names. Carolina duck, squealer, summer duck, woodie

Range. Breeds in forested parts of western North America from British Columbia south to California and east to Idaho, and in eastern North America from eastern North Dakota east to Nova Scotia, and south to Texas and Florida. Winters in the southern and coastal parts of the breeding range and southward uncommonly or occasionally to central Mexico, rarely to the Yucatan Peninsula. There is also a nonmigratory Cuban population.

Subspecies. None recognized.

Measurements. *Folded wing:* Delacour (1959): Males 250–285 mm, females 208–230 mm. Baldassarre (2014): Males, average of several studies, 208–240 mm, females 188–231 mm.
    Culmen (bill): Delacour (1959): Males 33–35 mm, females 30–33 mm. Baldassarre (2014): Males, average of several studies, 28–38 mm, females 31–38 mm.

Weights (mass). Nelson and Martin (1953): 248 males, average 1.5 lb. (680 g), maximum 2.0 lb. (906 g); 163 females, average 1.4 lb. (635 g), maximum 2.0 lb. Mumford (1954): 109 males, average 1.5 lb. (680 g); 99 females, average 1.44 lb. (652 g). Jahn and Hunt (1964): 49 fall adult males, average 1.6 lb. (725 g); 23 immature males, average 1.5 lb. (680 g).

Identification

In the hand. Male wood ducks, even in eclipse plumage, can be recognized in the hand by their iridescent upper wing surface and long, broad, and truncated tail, which is also somewhat glossy. Unlike all other North American duck species, both sexes have a silvery white sheen on the outer webs of the primary feathers and a bluish sheen near the tips of the inner webs.

In the field. Wood ducks sit lightly in the water, with their longish tails well above the surface. The birds are usually found not far from wooded cover. Often they perch on overhanging branches near shore and feed in fairly heavy woody cover that is flooded. The crest is evident on both sexes at a considerable distance, as is the male's white throat. The brilliant color pattern of males in nuptial plumage is unmistakable. In the air, wood ducks fly with great ease and apparent speed, the bill tilted below the axis of the body and the head often turned, giving a "rubber-necked" appearance, and the long tail is also evident. The underwing surface

Wood Duck *Aix sponsa*   33

The breeding (vertical hatching, with denser concentrations inked), wintering (shaded), and acquired or marginal (stippled) range of the wood duck.

is speckled with white and brownish, and the white on the trailing edge of the secondaries is usually visible, as is the white abdomen. The male has a clear whistle with rising inflection, and the female utters a somewhat catlike vocalization, but no true quacking notes.

Age and Sex Criteria

Sex determination. The tertial coverts of females are pinkish, while those of males are dark purple. Females also have large white "teardrop" tips on the secondaries, whereas males have narrow white tips on these feathers (Carney, 1964). In any adult plumage, the male's throat has two white extensions up the sides of the head, the eye is reddish to vermillion, and the bill is reddish at the base.

Age determination. In males, the tertials of juveniles are pale bronze, with pointed and frayed tips, while those of adults are deep purple, with blunt tips. These adult tertials grow in during the first fall of life. In immature birds the middle and greater coverts may show a mixture of the duller juvenal feathers and the very dark purple first winter coverts. In females, juveniles may have tertials that have pointed and frayed tips, rather than rounded tips, and the tertial coverts may be the greenish yellow of the juvenal plumage rather than the pink of the first winter plumage. In immature females the iridescent coloration usually does not extend onto the second row of middle coverts, and the most proximal greater covert of immatures is greener, duller, and smaller than those adjacent; in older females it is greener or lighter purple than those adjacent but of approximately the same size (Carney, 1964).

Distribution and Habitat

Breeding distribution and habitat. To a much greater extent than would be expected from a forest-adapted species, the wood duck in Canada is largely limited to the more southern regions. Its breeding range includes Vancouver Island, southern British Columbia, southern Alberta (locally), east-central and southeastern Saskatchewan, southern Manitoba (north approximately to The Pas), southwestern and southeastern Ontario, southern Quebec, and the Maritime Provinces (New Brunswick, Nova Scotia, and Prince Edward Island), plus Anticosti Island.

The United States range is divided into still-isolated eastern and western components, with a gap in the Rocky Mountain region and northwestern plains. The bulk of the North American breeding wood duck populations extends from the Missouri and Mississippi River valleys eastward over an area that more or less corresponds to the distribution of temperate deciduous and mixed deciduous-coniferous forests. The western limits of the eastern population have been slowly expanding since about the 1950s, as riverine forests of the Great Plains, no longer exposed to unlimited logging or periodic prairie fires, have matured and increasingly provided nesting cavities. Nest-box erection programs have supplemented the limited available natural cavities in this dry region. The western breeding limits currently (2016) occur in central Montana (Missouri and Yellowstone River valleys), central and southeastern Wyoming (Bighorn and North Platte River valleys), eastern Colorado (South Platte and Arkansas River valleys, west locally to Moffatt and Mesa Counties), western Oklahoma (Canadian and other river valleys, but west to Woods, Kiowa, and Love Counties), and west-central Texas (Rio Grande valley). The southern range limits have also been expanding in Texas, with breeding occurring uncommonly south to the Rio Grande. The relatively small wintering population in Mexico is reportedly increasing (Howell and Webb, 1995).

The western U.S. population's breeding range extends from Washington south to California, with the densest populations in northwestern Oregon and the central valley of California, and the eastern limits in northern and east-central Idaho and northwestern Montana. This population has been investigated rather little by comparison with the far larger eastern population.

The preferred summer habitat of wood ducks consists of freshwater areas such as the lower and slower-moving parts of rivers, bottomland sloughs, and ponds, especially where large willows, cottonwoods, and oaks are present (Grinnell and Miller, 1944). The presence of trees at least 16 inches in diameter (breast height), having cavities with entrances at least 3.5 inches wide and interiors at least eight inches in diameter, appear to be minimal nesting requirements (McGilvrey, 1968). Although cavities with extremely large entrances are rarely used, the height of the entrance and the depth of the cavity are not critical, nor is the direction of the entrance or its immediate proximity to water seemingly important (Grice and Rogers, 1965). However, the entrance should be protected from weather, and the cavity must be well drained.

Besides the presence of usable nesting sites, the breeding habitat must contain adequate food sources, suitable cover, available water, and suitable brood-rearing locations. McGilvrey's summary of these requirements indicates that foods should include overwintering seeds or nuts (acorns, domestic grains, etc.), native herbaceous plants, and aquatic or aerial insect life. Breeding cover should include trees, shrubs, or both. The trees should have low branches, providing overhead and lateral cover, and preferably should be flooded. Shrubs that have strong stems rising and spreading out about two feet above the water level, such as buttonbush (*Cephalanthus*), are highly desirable.

For best foraging the water should be no more than 18 inches deep, still or only slow-moving, and available throughout the incubation period. Ideal brood-rearing habitat includes a source of available foods for ducklings (such as insects and duckweeds), water persisting throughout the fledging period, and dense overhead cover, such as provided by flooded shrubs or dead tree tangles. The presence of herbaceous aquatic plants is highly desirable, as are resting sites for the brood.

Population. In the middle of the twentieth century, Benson and Bellrose (1964) estimated that about half of a continental population of 400,000 breeding pairs in 1962 bred in the northern halves of the Atlantic and Mississippi Flyways. Sincock et al. (1964) believed that the 12 states in the southern halves of these flyways might produce about 650,000 wood ducks annually. Naylor (1960) estimated that of a total western breeding population of about 16,000 pairs in 1958, 7,500 were in Oregon, 6,000 were in Washington, 1,500 were in California, and the remaining 860 were located in Idaho, British Columbia, and Montana.

Wood duck populations have increased greatly since the 1960s. Bellrose and Holm (1994) judged that Atlantic Flyway populations increased 10.3 percent annually between 1959 and 1990, and Mississippi

Flyway populations 8.8 percent. Population estimates during the early 2000s include 2,800,000 birds for eastern North America, 665,000 for central regions, and 66,000 for western regions (Wetlands International, 2006). The estimated hunter-kill estimate in the United States in 2014 was about 330,000 birds, and in Canada was about 62,000 birds.

The woodland breeding habitats of wood ducks makes national aerial surveys impossible, but breeding population estimates for the eastern segment of the wood duck's U.S. range by Bellrose and Holm (1994) included about 1.07 million birds in the Atlantic Flyway, with 230,000 in Ontario, 104,000 in Maryland, 91,000 in Virginia, 87,000 in New York, 84,000 in North Carolina, 72,000 in South Carolina, 55,000 in Georgia, and 33,000 in Florida.

In the upper Midwest about 2.4 million breeding birds were estimated by Bellrose and Holm (1994) to be present in the early 1990s, with large numbers in Minnesota (342,000), Wisconsin (237,000), and Michigan (140,000). Great Plains populations were estimated by them at 13,000 in North Dakota, 10,000 in Oklahoma, about 9,000 in South Dakota, and 4,000 to 5,000 in Nebraska. In the Gulf coast region, the estimated population was 242,000 birds, with about half in Louisiana, about 30 percent in Mississippi, and 15 percent in Alabama.

Estimates by Bellrose and Holm (1994) of the west coast segment of the wood duck's range from 1982 to 1987 included an average total of about 60,000 birds, with the largest numbers in California (26,000), followed by Oregon (17,000), Washington (7,900), and British Columbia (3,000).

Canada's eastern breeding wood duck population was estimated by Dennis (1990) as averaging 117,000 from 1972 to 1985, with 79 percent in Ontario, 16 percent in Quebec, and 2 percent in New Brunswick. The regional total estimate is far below Bellrose and Holm's Canadian estimate noted above, showing the problems of large-scale censusing of wood ducks.

Wintering distribution and habitat. Virtually the entire North American wood duck population winters within the borders of the United States; a few overwinter in southwestern and southeastern British Columbia, and in northern Mexico the wood duck is an uncommon winter migrant. The western population of wood ducks winters primarily in the Central Valley region of California; Naylor (1960) reported that California supported most of an estimated wintering population of about 55,000 birds. No more recent estimates are available; the scattered distribution and forest-related habitat of wood ducks makes winter population counts both difficult and unreliable.

The eastern wood duck population is many times larger than the western one but in recent years has been largely overlooked during midwinter surveys. Counts made in the early 1960s indicated about 100,000 birds wintering in the Mississippi Flyway and progressively smaller numbers in the Atlantic and Central Flyways. Bellrose and Holm (1994) recognized three wintering zones for wood ducks in the Atlantic Flyway: the little used northern zone (north to southern Ontario), the primary wintering area in the central zone (Virginia and North Carolina), and the southern zone (South Carolina and Florida). Frank Bellrose (cited by Baldassarre, 2014) estimated that 299,000 wood ducks annually winter in South Carolina, 267,000 in North Carolina, 243,000 in Georgia, and 167,000 in Florida. Similar estimates by Bellrose for the Gulf coast region were 338,000 for Mississippi, 321,000 for Arkansas, 188,000 for Louisiana, and 184,000 for Alabama.

Fig. 5. Male and female wood ducks.

Recoveries of wood ducks banded in Wisconsin indicate that these birds move south along the Missis-sippi River valley to winter in Arkansas, Louisiana, Texas, and Mississippi and move farther east only to a limited extent (Jahn and Hunt, 1964). On the other hand, wood ducks banded in Massachusetts evidently move south along the Atlantic coastal plain and winter primarily in the Carolinas, Georgia, and northern Florida; only a few recoveries are found as far west as Louisiana and Mississippi (Grice and Rogers, 1965). It would thus seem that the Mississippi River and its tributaries provide one major migratory thorough-fare, and the Atlantic coast another, with uplands and mountains being avoided and producing barriers to population interchange. This was confirmed by Bellrose and Holm (1994), who found that birds migrat-ing from the northeastern states tended to move west after passing the Appalachian Mountains, from the Florida panhandle to as far west as eastern Texas.

Secluded freshwater swamps and marshes are the favored wintering habitats of wood ducks throughout the southern states, particularly where acorns, hickory nuts, water-lily seeds, and similar foods are read-ily available. Stewart (1962) noted that fall migrant wood ducks congregate where the masts of beech and oaks are available, and they also utilize interior impoundments with stands of spatterdock (*Nuphar*). Small numbers use fresh estuarine bay marshes, especially where narrowleaf cattail (*Typha augustifolia*) and white water lily (*Nymphaea odorata*) are present. Among the estuarine river marshes, the largest spring and fall populations are found in fresh or slightly brackish water, especially where arrow arum (*Peltandra*) is common. Flooded forests are favored wintering habitats, especially those with abundant invertebrate foods.

General Biology

Age at maturity. A one-year period to maturity is well established for wood ducks. Ferguson (1966) noted that 19 of 24 aviculturists reported breeding by captive birds in the first year, while the remainder reported second-year breeding. As summarized by Grice and Rogers (1965), many studies have reported that birds marked as juveniles often returned to the same area the following year for nesting. Among 95 wild females tagged by Grice and Rogers, 30 were known to be nesting as yearlings. Since many of the marked birds were not located, the actual percentage of nesting by wild yearling females was certainly higher.

Pair-bond pattern. Apparently pair-bonds are renewed yearly, since males normally desert females at the beginning of incubation and the females rear their young alone (Grice and Rogers, 1965). On occasion, however, males have been seen in company with females and broods, and there is at least one record of a male incubating (Rollin, 1957).

Nest location. A number of studies on natural nesting cavities of wood ducks have been made, and several general characteristics of cavity requirements have emerged. McGilvrey (1968) summarized the optimum natural cavity as having a height of 20 to 50 feet, an entrance 4 inches in diameter, a cavity bottom of 100 square inches, a cavity depth of 24 inches, and a tree diameter of 24 to 36 inches. There appears to be a preference for high cavities and those with small entrances, which raccoons are unlikely to be able to enter (Bellrose et al., 1964; Weier, 1966). Apparently there is also a preference for nesting in rows or clusters of large trees of similar size, rather than in isolated large trees (Grice and Rogers, 1965). Open tree stands are also preferred over dense woods. At least in the case of artificial cavities (nest boxes), those situated over water are greatly preferred to those on land. Cavities with entrances only slightly larger than the minimum possible (3.5-by-4 inches) are preferred, as are those with cavity depths of less than 50 inches (Bellrose et al., 1964).

Clutch size. Estimates of clutch size are often confused by dump-nesting involving several females, which tends to inflate estimates of clutch size. Naylor (1960) estimated that 13.8 eggs represented a normal complete clutch, while dump-nests averaged 28.5 eggs per nest. Similarly, Cunningham (1969) noted that the average clutch size of "single" nests ranged from 13.5 to 15.9 during three years, while that of dump-nests averaged about 28 eggs. The incidence of dump-nesting was related to population density. Leopold (1966) reported an average clutch of 13.9 eggs for early nests. He noted that of 297 potential "egg days," only 13 were missed; thus the egg-laying rate averaged 1.04 days per egg. Renests usually average smaller clutches (Leopold, 1966), and as many as two renesting attempts have been noted (Grice and Rogers, 1965).

Wood duck, adult male

Many instances of double brooding have been found since the 1960s, especially in the southern states, where breeding seasons are longer, but even as far north as Missouri (Rogers and Hansen, 1967), where 3.8 percent of all the broods found were second broods (Fredrickson and Hansen, 1983). Moorman and Baldassarre (1988) reported that at Eufala National Wildlife Refuge in Alabama 7 of 101 successful nests resulted from second broods in 1985, and 16 of 133 broods the following year. The females that produced second broods averaged 2.6 years of age, and the clutch sizes of the second nestings were lower than numbers in the first nests. During the two years of study the elapsed time between the hatching of the first clutch and the initiation of the second nesting ranged from 15 to 72 days. In the San Joaquin Valley of California, 56 second broods were found among 1,540 total nesting efforts (Thompson and Simmons, 1990), and in South Carolina 21 of 219 successful nesting efforts were the result of second efforts (Kennamer and Hepp, 1987).

Brood care in wood ducks typically lasts 30 to 40 days, so in southern regions the approximate 120 to 140 days needed to hatch and rear two broods to independence could be easily fulfilled. Although at one time ruddy ducks were believed to produce second broods (Kortright, 1942), wood ducks are the only North American duck species so far known to do so, but muscovies might be another possibility, given their long available breeding seasons in the tropics.

Incubation period. The incubation period averages about 30 days, with reported extremes of 25 to 37 days (Grice and Rogers, 1965). Leopold (1966) noted that about half the clutches hatch in 30 days and two-thirds in 29 to 31 days, with pipping starting two days prior to hatching.

Fledging period. Grice and Rogers (1965) noted that about 70 percent of the juveniles he studied were capable of flight (after being thrown into the air) at 60 days of age, before their primaries were fully grown (ducks can usually begin to fly when their primaries have reached 80 percent of their ultimate length).

Nest and egg losses. A large number of studies of wood duck nests have been made, and most indicate fairly high success rates, especially when nesting boxes are used. Weller (1964) summarized three studies (mostly using artificial nesting boxes) that totaled 1,648 nests and averaged a 66 percent nest success. Leopold (1966) reported a 94 percent nesting success for 281 nests, and a 75 percent hatching success for 2,860 eggs. In the majority of studies, the single most important predator was the raccoon, and by the construction of relatively raccoon-proof nesting boxes, the nesting success was generally quite high (Grice and Rogers, 1965). In areas where European starling populations are high, 20 percent or more of the nests have sometimes been destroyed, but the starlings' choice of wood duck nesting boxes can be reduced by constructing houses that are too well lighted for the light-intolerant starlings (Bellrose and McGilvrey, 1966).

Bellrose and Holm (1994) reported that over a long-term study the most important sources of nest loss in decreasing significance were raccoons, fox squirrels, European starlings, bullsnakes, and woodpeckers. In Missouri the corresponding relative significance of nest loss causes were black rat snakes, desertion, raccoons, and starlings (Hansen, 1971). In the southern states various snakes are known to be important, and locally or occasionally fox squirrels, minks, opossums, or rats also pose problems for nesting birds. Duckling predators include minks, turtles, fish, snakes, bullfrogs, and various other birds and mammals (McGilvrey, 1968).

Juvenile mortality. Grice and Rogers (1965) determined that of 135 broods studied over a three-year period, brood size was reduced from an average of 12.5 at hatching to 5.8 at the time of fledging, or a loss of approximately 50 percent of the young during the fledging period. Grice and Rogers found that early-hatched broods had the lowest mortality, while late-hatched young had an average brood size of 9.9 at hatching and only 2.2 at fledging. Jahn and Hunt (1964) also calculated a collective average brood size of 5.8 young for birds near fledging, based on six different field studies. Estimates of first-year mortality rates for birds banded as juveniles range from 61.7 to 82.5 percent, with an average of three New England studies being 76.7 percent (Grice and Rogers, 1965).

Wood duck, adult male (foreground) courting female (background)

Adult mortality. Studies of banded birds in three New England states have provided estimated annual adult mortality rates of 51.7 to 63.7 percent, with an average of 58.9 percent (Grice and Rogers, 1965). Nichols and Johnson (1990) used banding data from six different areas to estimate collective adult male survival rates of 55.6 percent for adult males, 50.6 percent for adult females, 47.5 percent for immature males, and 43.2 percent for immature females.

General Ecology

Food and foraging. A considerable number of food analyses (Martin, 1951) of wood ducks have consistently pointed toward a high usage of fruits and nuts of woody plants, such as dogwood and elm trees, including beechnuts, acorns, hickory nuts, as well as a substantial consumption of the seeds of floating-leaf aquatic plants (*Brasenia*, *Numphaea*, *Nuphar*). Additionally, the seeds and vegetative parts of other aquatic plants such as wild rice (*Zizania*), pondweeds (*Potamogeton*), arrow arum (*Peltandra*), duckweeds (*Lemna* and others), and bur reed (*Sparganium*) are consumed in large quantities. Stewart (1962) found that in the Chesapeake Bay area wood ducks feeding on river bottomlands consumed mostly beechnuts and acorns, while birds in the estuarine river marshes predominantly consumed the seeds of arrow arum.

Oak species that produce fairly small acorns are used by wood ducks more often than those that produce large acorns, particularly oaks in bottomland soils that are occasionally flooded (Brakhage, 1966). The former include such species as pin oak (*Quercus palustris*), water oak (*Q. nigra*), willow oak (*Q. phelios*), and Nuttall oak (*Q. nuttallii*). Wood ducks may search for acorns among the forest litter or sometimes pluck them from the branches before they have fallen. When foraging on water they tip-up but only rarely dive for food. In one study only female wood ducks were observed diving while foraging (Kear and Johnsgard, 1968), but both sexes have been seen diving for acorns in water up to about ten feet in depth (Briggs, 1978). Preferred foraging habitat consists of water that is no more than 18 inches deep, the approximate limit a wood duck can reach by tipping-up.

Sociality, densities, territoriality. During most of the year, the wood duck is found only in small flocks of a dozen birds or less, with larger aggregations occurring only during the nocturnal roosting period. Both on the wintering grounds and during migration such social roosting is typical, and roosts sometimes support hundreds of birds. Hester (1966) noted that roosts vary in size from less than an acre to several acres, and the numbers of birds using them range from less than a hundred to several thousand, including one recorded roost of 5,400 birds.

On arrival at their nesting grounds, wood ducks are usually in small groups of up to a dozen birds, and usually already in pairs. Once established on their nesting areas, pairs do not seem to restrict their movements to a particular territory or defend an area as such, but rather the males simply protect their females from attentions by other males (Grice and Rogers, 1965).

Breeding densities are apparently determined by the availability of suitable nesting cavities, which are usually fairly limited unless supplemented by artificial nesting boxes. In one study, at a site where boxes were not used, 37 of 67 cavities on 442 acres were used during one year (Bellrose et al., 1964), for a density of about 12 acres per nest. Examples of high nesting densities achieved with nesting boxes include 41 nests on one eight-acre pond, 95 nests on a 150-acre refuge, and 37 nests on 100 acres (McGilvrey, 1968).

In an earlier book (Johnsgard, 1997), brood parasitism among birds worldwide was discussed. Of the approximately 150 species of Anatidae, intraspecific brood parasitism has been observed in at least 64 species. Of these, the wood duck has been the subject of the greatest number of published studies and has had the greatest reported range (23–95 percent) of parasitism frequency (the percentage of available nests parasitized). Reported hatching success rates for wood duck eggs in parasitized nests (including dump-nests) among five studies ranged from 39 to 81 percent, and in nonparasitized nests the rates ranged from 47 to 87 percent, suggesting that there was not a large reproductive penalty for being exposed to intraspecific brood parasitism.

Interspecific relationships. Because of their specialized nesting adaptations, competition for nest sites between wood ducks and other duck species is limited. The common goldeneye has an overlapping breeding range, but this occurs only near the northern edge of the wood duck's range. A study in New Brunswick (Prince, 1968) indicated that competition between the two species was limited because of their differing nest-site and cavity-size preferences as well as differences in their preferred foraging and loafing areas. Wood ducks also used areas with somewhat larger trees. Cavities used by the two species were similar in their entrance sizes, but goldeneyes evidently preferred cavities that were less deep and of a more consistent inside diameter as compared with those preferred by wood ducks.

In many states and Canadian provinces, hooded mergansers and wood ducks pose a brood parasitism problem for each other (Hansen, 1971; Robinson, 1983; Zicus, 1990; Mallory et al., 2002). Mallory et al. (2002) reported that in Ontario 11.3 percent of 227 hooded merganser nests contained wood duck eggs, and in Missouri Hansen (1971) found 15 mixed clutches of these species. In Minnesota 5 of 13 wood duck nests studied by Zicus (1990) included one or more hooded merganser eggs, and hooded merganser eggs represented 12 percent of all the eggs that were incubated by wood ducks.

At the southern end of their breeding range in Texas, wood ducks overlap with black-bellied whistling ducks (*Dendrocygna autumnalis*), which often leads to competition for nesting sites (Bolen and Cain, 1968; Labuda, 1969). Competition for nesting cavities may also occur regionally with other species. McGilvrey (1968) noted that nest-site competitors of wood ducks include the European starling, American kestrel, and various screech-owls, squirrels, bees, and hornets. Squirrels can be serious competitors for nests in some areas, especially where only natural cavities are available.

General activity patterns and movements. The evening roosting behavior of wood ducks is well known and has been frequently studied as a population index technique. These flights are usually most pronounced during fall and winter. A study by Martin and Haugen (1960) indicated that the morning flights lasted for about 45 minutes and usually ended by15 minutes after sunrise. Early evening flight activity mainly occurred during the last 50 minutes before sunset, but both morning and evening flights gradually occurred nearer the periods of darkness and were made during a shorter period of time as the fall season progressed.

Using color-banded birds, Stewart (1958) studied local movements of broods and families. He found that at the age of about two weeks, broods moved away from their natal sites into new habitats and often merged with other wood ducks. Some of such brood movements were quite long, with a maximum record movement of 3.5 miles. When leading broods, females continued to make their morning and evening feeding flights and started gathering into small groups when the ducklings were about six weeks old. At the age of eight weeks, when the young have fledged, additional congregation occurred, with some segregation of adult and young birds. In early October the ducks moved from ponds and lakes to rivers and creeks, usually within 15 miles, and by late October the fall migration had begun.

Social and Sexual Behavior

Flocking behavior. Social nocturnal roosting behavior is common in wood ducks, with up to as many as 3,000 birds having been reported from various areas. Roosting sites are usually close to the habitats that are used during daytime and often consist of flooded trees or wooded swamps (Bellrose and Holm, 1994). Stewart (1958) noted that in late-summer concentrations, the morning flights away from the roosting sites consisted of larger flocks than did the evening flights back to the roost, which usually numbered from 1 to 20 birds.

Pair-forming behavior. Pair formation evidently occurs on the wintering grounds, since birds arrive at their nesting areas already paired (Grice and Rogers, 1965). The pair-forming displays of wood ducks are numerous and complex (Johnsgard, 1965, see Fig. 6), but an integral feature of pair formation is the performance of inciting by a female toward a specific male. In effect, the female incites a particular male to attack other birds, usually other males. This inciting behavior is highly ritualized and rarely leads to attacks. Instead, the "preferred" male responds to inciting by swimming ahead of the female and turning the back of his head toward her. This combination of inciting and turning-of-the-back-of-the-head display seems to be a fundamental feature of pair formation in nearly all true ducks (Johnsgard, 1960).

Copulatory behavior. Unlike other North American surface-feeding ducks, copulation in wood ducks is preceded by the female assuming a prone position well in advance of treading. I have seen no preliminary mutual displays by the pair prior to the female's assumption of this posture, in which she lies flat on the water with her head low and her tail tilted slightly upward. The male typically swims around her, making drinking or bill-dipping movements and sometimes pecking gently at her. Mounting then occurs, and after treading is completed, the male usually first swims rapidly away from her while turning-the-back-of-the-head and then turns and faces the bathing female (Johnsgard, 1965).

Nesting and brooding behavior. Leopold (1966) reported that mated pairs begin to look for nests shortly after they arrive in late March, spending several mornings investigating possible sites. The male accompanies the female but does not enter the nesting box. After five to six days of such behavior, the first egg is laid. Egg-laying occurs in early morning while the mate waits nearby, and afterward the birds leave until the following morning. Down-picking begins with the fourth to eighth egg. While the last few eggs are being laid, the female may spend the night in the box, presumably picking down. Incubation begins with the last egg, and during the incubation period two rest periods are normally taken daily, during early morning and late afternoon hours. The male usually accompanies the hen on such flights, until he deserts her for his postnuptial molt. During first-time nestings, the male usually attends the female into the fourth week of incubation. The female remains in the nest during the four- to six-hour hatching period, and the family usually spends its first night in the nest. The next morning the female usually takes her rest flight and then returns to the nest and calls the young from the cavity with a series of low *kuk* notes. The young jump from the nest in rapid succession, and the family then walks to the nearest water.

Fig. 6. Sexual behavior of wood duck (A–F) and gadwall (G–H), including (A–B) display shake, (C–D) bill-jerking, (E) postcopulatory facing the female, (F) chin-lifting, (G) grunt-whistle, and (H) turning-of-the-back-of-the-head.

Stewart (1958) noted that newly hatched broods went to water areas that were nearest the hatching place, provided that vegetative cover was present. For the first two weeks of life little brood congregation occurs, although lost individual ducklings may attach themselves to other broods. Because of such brood merger, age differentials among ducklings in broods are not uncommon.

Postbreeding behavior. Following their desertion of the females, male wood ducks evidently move to secluded woodland ponds or swamps, where they are rarely seen. Females undergo their molt later than males; they probably normally leave their broods and begin to molt between six to eight weeks after the young have hatched. Like the males, they then inhabit the thickest possible cover and are almost never seen (Grice and Rogers, 1965). Shortly after regaining flight, the young and the adults begin to congregate in larger flocks in preparation for their fall migration.

Tribe Anatini
(Dabbling or Surface-feeding Ducks)

Eurasian Wigeon
Anas penelope Linnaeus 1758

Other vernacular names. European wigeon, wigeon (UK)

Range. Breeds in Iceland and temperate portions of Europe and Asia south to England, Germany, Poland, and east through Turkmenistan, northern China, and northwestern Mongolia to the Pacific coast and Kamchatka. Winters in Europe, northern and central Africa, and Asia. Regularly seen from fall through spring in North America, especially along the Atlantic and Pacific coasts, and occasionally in the continental interior. Not yet proven to be a breeding species in North America, but occasional breeding seems highly probable, and several wild hybrids have been documented.

Subspecies. None recognized.

Measurements. *Folded wing:* Delacour (1956): Males 254–270 mm, females 236–255 mm. Owen (1977): Adult males average 269.2 mm, females 247.1 mm. Kear (2005): 271 males, average 267 mm; 483 females, average 250 mm.
   Culmen (bill): Delacour (1956): Males 33–36 mm, females 31–34 mm. Owen (1977): Adult males average 35.1 mm, females 33.1 mm.

Weights (mass). Schiøler (1925): 42 adult males (Greenland), average 819 g (1.81 lb.), 23 immature males, average 706 g (1.56 lb.), maximum 1,073 g; 24 adult females (Greenland), average 724 g (1.6 lb.), 20 immature females, average 632.5 g (1.39 lb.), maximum, 962 g. Owen (1977): adult males average 721 g, females 662 g.

Identification

In the hand. Either sex may be safely distinguished in the hand from the American wigeon by the presence of dark mottling on the underwing surface, particularly the axillaries. Eurasian wigeons may be distinguished from other surface-feeding ducks by the white to grayish upper wing coverts and the green speculum pattern, with a black anterior border. Both sexes are more brownish on the cheeks and neck than is true of the American wigeon.

Males in breeding plumage have a cinnamon-red head and neck with a buff crown and forehead and sometimes a trace of iridescent green behind the eyes (thus a greenish postocular stripe is not proof of American wigeon hybridization). The upper breast is purplish pink, and the lower breast, abdomen, flanks, and mantle are white, mostly finely vermiculated with dusky coloration. The rump is light gray and the tail coverts are black, except for the middle upper coverts, which are whitish. The tail is gray to brownish centrally and white to silvery gray outwardly. The upper wing-coverts are mostly white, except for the secondary coverts, which are tipped with black. The secondaries are iridescent green with blackish tips; the tertials are black, edged with white; and the primaries and their coverts are ashy brown. The axillaries are mottled or flecked with brown or grayish coloration. The iris is brown, the bill pale bluish gray with a black tip, and the legs and feet gray.

Males in eclipse closely resemble females but have white upper wing-coverts.

Females have a cinnamon buff head and neck (grayish buff in some birds), flecked with dusky or greenish coloration. The upper breast and sides are buffy or reddish brown (grayish brown in some), marked with dusky coloration. The scapulars and rump are dusky brown, the longer scapulars being edged with buff or white. The upper wing-coverts are mostly dusky gray or brown, with whitish tips, while the greater secondary coverts are tipped with black and white. The secondaries have a dull green to blackish speculum, with a narrow terminal bar. The axillaries are gray, mottled with dusky coloration. The tail and the soft-part colors are as in the male.

Juveniles resemble adult females, but males gradually assume nuptial plumage during their first fall. However, young males retain grayish brown upper wing-coverts during their first winter, and first-winter females have less obvious white wing-covert tips than do adult females.

In the field. Females are not considered safely separable from the female American wigeon in the field, but if both species are together the more brownish and less grayish tones of the Eurasian species will be evident. Males in nuptial plumage are easily recognizable, since they exhibit a creamy yellow rather than a white forehead, and a cinnamon-red head and neck color instead of a light grayish one. Since some male Eurasian wigeon exhibit a green iridescence around and behind the eye, similar to that of the American wigeon, this is not a reliable field mark for distinguishing the two. The call of the male Eurasian wigeon is a shrill double whistle, sounding like *whee-uw*, while that of the American species is a series of weaker repeated single notes. Calls of the females are nearly identical. In flight, the mottled under wing-coverts and axillaries might be visible under favorable conditions.

Fig. 7. Male Eurasian wigeon.

Age and Sex Criteria

Sex and age determination. Probably the same sex and age determination criteria as indicated for the American wigeon apply to this species.

Occurrence in North America

The great number of specimen and visual records of Eurasian wigeon in North America has led several people to speculate that breeding, of at least a local or periodic nature, must occur on this continent. Hasbrouck (1944) compiled nearly 600 North American sight or specimen records for this species through the early 1940s. On the basis of these he concluded that a regular southward fall migration occurs along the Atlantic and Pacific coasts, followed by an apparent northward spring migration through the continental interior. Of the records he presented, about 60 percent are from states or provinces largely or wholly in the Atlantic Flyway. The remainder was about equally divided among the states and provinces representing the Pacific and Mississippi Flyways, while only about 2 percent of the records were from Central Flyway states.

The Pacific Flyway states and provinces for which Hasbrouck listed records extended unbroken from Alaska to California. Hasbrouck listed records from the Central Flyway states of Wyoming, Nebraska, and Texas; records for all the Mississippi Flyway states except Tennessee, Arkansas, Mississippi, and Alabama; and for all the Atlantic Flyway states except Vermont and West Virginia.

A later summary by Edgell (1984) for the years 1948 to 1981 totaled about 1,500 birds seen during fall and winter, and 560 during spring, of which 55 percent came from the Pacific Flyway, mostly California and Washington. The spring and summer records were more scattered, with 39 percent from the Pacific Flyway. Sightings increased greatly after the mid-1960s, and Edgell judged that by 1981 about 1,000 Eurasian wigeons were wintering in North America. Since the appearance of eBird online, there have been sight records for all 49 of the continental states, all the Canadian provinces, and two of the three Canadian territories (Nunavut was the exception as of 2016).

Apparently California is a major wintering area for Eurasian wigeons, especially the Sacramento Valley, where they are frequently seen at national wildlife refuges, state wildlife areas, flooded rice fields, and other regional wetlands. It is believed that many of these birds use the Klamath Basin as a prime spring stopover area, based on telemetry data from the USGS Western Ecological Research Center (http://www.werc.usgs.gov/Project.aspx?ProjectID=43). In Oregon it is most common near Portland, in the lower Columbia River valley, and along the coast (Gilligan et al. 1994).

Although Texas is well off the coastal migratory corridors of the Eurasian wigeon, 53 documented records of this species were made in Texas between 1988 and 2014. All were reported between October 3 and May 8, including a few apparent American × Eurasian wigeon hybrids (Lockwood and Freeman, 2014).

During US hunting seasons from the mid-1990s to 2008 a maximum single-season estimate of 190 Eurasian wigeons were killed in the Atlantic Flyway, and a maximum of 2,120 in the Pacific Flyway. Total US hunter-kills have averaged about 1,200 annually since 1994, but Eurasian wigeons were apparently not distinguished from American wigeons during earlier US hunter-kill surveys. Estimated recent total annual Canadian kills have ranged from about 50 to 750 birds. The total 2014 US kill of American wigeons was about 611,000 birds, and in Canada was about 13,000. Thus, Eurasian wigeons have made up about 0.002 of all wigeons identified between the total US and Canadian kills.

Eurasian wigeon, adult male

In spite of all these recent occurrences, including summer sightings of pairs and many wild hybrids (e.g., Merrifield, 1993), there is still no definite evidence of Eurasian wigeons breeding in North America. This might be occurring in remote areas of Alaska or perhaps in Canada's Yukon or Northwest Territories.

American Wigeon
Anas americana Gmelin 1789

Other vernacular names. Baldpate, poacher, whistler, wigeon

Range. Breeds in northwestern North America from western Alaska, Yukon Territory, and Northwest Territories east to southern Nunavut, Hudson Bay, southern Quebec, and the Maritime Provinces, also south to California, Utah, Colorado, Nebraska, and the Dakotas, and east to Michigan, locally to New York and Maine. Wintering occurs along the Pacific coast from southern Alaska south to southern Mexico (infrequently south along the Pacific slope of Central America to Panama), the southern United States and Gulf coast, and along the Atlantic coast from southern New England south to Florida and the West Indies.

Subspecies. None recognized.

Measurements. *Folded wing:* Phillips (1924): Males 252–270 mm, females 236–258 mm. Kear (2005): Males (12) average 264 mm; females (12) average 246 mm. Owen (1977): Adult males average 259.2 mm, adult females average 247.1 mm.

   Culmen (bill): Phillips (1924): Males 45–48 mm, females 33–37 mm. Kear (2005): Males (12) average 37 mm; females (12) average 36 mm. Owen (1977): Adult males average 35.1 mm, adult females average 33.1 mm.

Weights (mass). Nelson and Martin (1953): 264 males, average 1.7 lb. (770 g), maximum 2.5 lb. (1,032 g); 108 females, average 1.5 lb. (680 g), maximum 1.9 lb. (861 g). Jahn and Hunt (1964): 29 fall adult males, average 2 lb. (907 g); 173 immature males, average 1.94 lb. (879 g); 28 adult females, average 1.94 lb. (879 g); 146 immature females, average 1.69 lb. (765 g).

Identification

In the hand. Apart from the rare Eurasian wigeon, American wigeons are the only North American surface-feeding ducks that have white or nearly white upper wing-coverts, separated from a green speculum by a narrow black band. The rather short bluish bill and similarly colored legs and feet are also distinctive; only the northern pintail has comparable bill and foot coloration, and this species lacks pale gray or white on the upper wing-coverts. See the Eurasian wigeon account for distinction from that species.

In the field. American wigeon can be recognized on land or water by their grayish brown to pinkish body coloration. They often feed on land, eating green leafy vegetation, and float about buoyantly in shallow

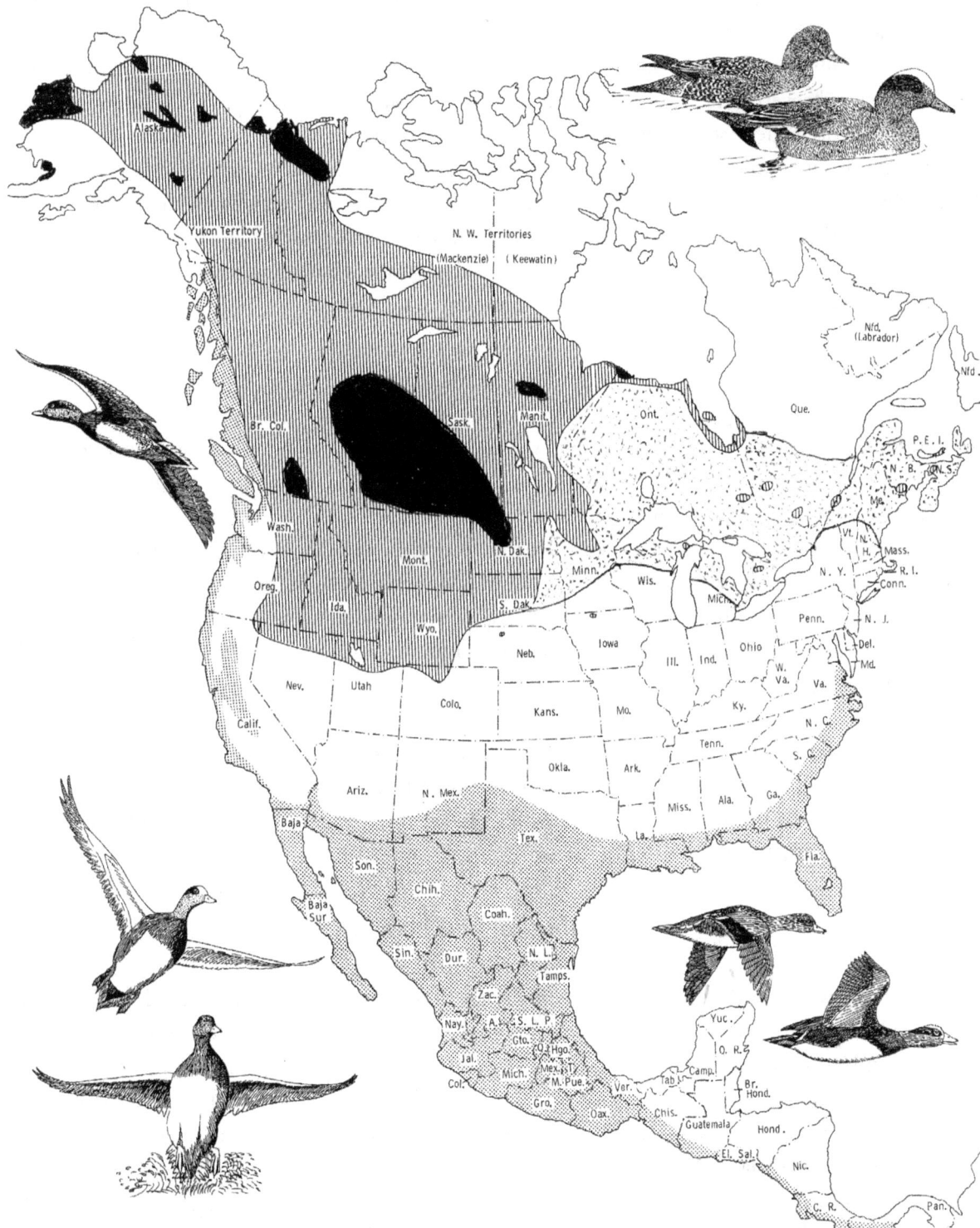

The breeding (vertical hatching, with denser concentrations inked), wintering (shaded), and acquired or marginal (stippled) range of the American wigeon.

water, where they feed on aquatic leafy materials or steal it from diving ducks. The short bill and similarly short, rounded head are often evident, and when the male is in nuptial plumage his pure white forehead markings are visible for great distances, as are the large white areas on the sides of the rump, contrasting with the black tail coverts. The white upper wing-coverts are usually not visible when the bird is at rest, but when in flight this is the best field mark, alternately flashing with the grayish underwing surface and with the white abdomen of both sexes. American wigeon are about the same size as gadwalls and often mix with them in flight. Both species have white underparts, but while the gadwall exhibits white at the rear of the wing only, the wigeon exhibits dark secondaries and white on the forward half. Males often call in flight or when on the water, uttering a repeated and rather weak whistle. Females are relatively silent ducks, and their infrequent, guttural quacking notes are not repeated in long series.

Age and Sex Criteria

Sex determination. *Adult males* also have long, sharply pointed tertials that are black on the outer web and have narrow white margins, and the greater tertial coverts of adult males are gray. Vermiculations on the scapulars, back, or sides and entirely white middle coverts indicate an adult male. *Females* have shorter tertials that are brownish gray edged with white on the outer web, and the greater tertial coverts of adult females are dark brown with white edges. *Immatures* may be sexed by their middle coverts, which in males vary from dirty white to dark, with light centers surrounded by poorly defined cream or gray edging, and in immature females are dark, without light centers and usually with fairly well-defined light brown edging (Carney, 1964).

Age determination. Immatures of both sexes have small, light-edged, and brownish tertials that are often faded or frayed. The greater tertial coverts may also be frayed and faded (Carney, 1964). The tail feathers may also have notched tips until they are molted in fall or winter. See also Esler and Grand (1994) for age determination in spring birds.

Distribution and Habitat

Breeding distribution and habitat. The breeding range of the American wigeon is broad, extending from the Bering coast to the Atlantic, and from the Beaufort Sea coast and Hudson Bay south to northeastern California and the northern parts of Utah, Colorado, and Nebraska. The range map's dashed lines indicate apparently expanded breeding regions since the 1970s, and the dotted lines indicate the northern limits of recent wintering regions. As a result of global warming, most other American waterfowl are also now wintering at more northerly latitudes than was the case during the 1970s, a phenomenon that is especially apparent in the Great Plains (Johnsgard, 2009).

Studies on breeding habitat preferences are limited. Keith (1961) compared the percentage of paired ducks on three different lake areas and two areas of potholes in southeastern Alberta. He noted that each of

Fig. 8. Male American wigeon.

the three lake areas accounted for more than 20 percent usage by wigeons, while the two areas of potholes had 10 to 20 percent usage. The highest usage (nearly 30 percent) occurred on a large 20.8-acre lake with an average depth of 3 to 3.5 feet, limited emergent vegetation, and a relatively large amount of water mil-foil (*Myriophyllum*) and pondweeds (*Potamogeton*) among the submerged plants. Potholes received even less relative use by broods, whereas the lake just mentioned accounted for about 40 percent of the brood use. Gadwalls exhibited a similar pattern of habitat use by pairs and broods.

Munro (1949) noted that wigeon prefer to nest around certain lakes or marshy sloughs that are sur-rounded by dry *Carex* meadows, in which the nests are placed. Unlike most dabbling ducks, females and young frequent the open water of marshy ponds, lake bays, or marsh-edged rivers, with this preference for open water perhaps related to the commensal foraging relationship between wigeons and diving waterfowl. The closely related Eurasian wigeon likewise prefers to nest where shoreline meadow belts are present, and additionally apparently requires partly wooded shorelines, since it is absent from both open tundra and small forest ponds (Hildén, 1964). To some extent, the American wigeon also shows a tendency to nest in open wooded or brushy habitats (Phillips, 1924).

Population. North American breeding grounds surveys in 2015 indicated a total population of 3.0 million birds, 17 percent above the long-term average of 2.6 million (USFWS, 2016). Total US hunter kills averaged about 625,000 in 2014 and 2015, with no clear long-term directional trend apparent. The estimated total annual Canadian kill in 2014 was 38,000, and during the early 2000s had similarly ranged from about 30,000 to 50,000.

Wintering distribution and habitat. The wintering distribution of this wigeon is extensive, with nearly half of the winter population occurring in the Pacific Flyway, judging from census figures of the 1970s, and close to 60 percent in surveys from the early 2000s. Large numbers of birds (nearly 30 percent) wintered in the Central Flyway, but relatively few winter in the Mississippi or Atlantic Flyways. In the western states, wigeon occur from Puget Sound to Oregon's Willamette Valley and south along California's Humboldt Bay, other coastal bays, rivers, and inland valleys, pastures, and wet meadows where green vegetation is readily available (Chattin, 1964), but the largest number (perhaps 30 to 40 percent of the total US population, and about 70 percent of the Pacific Flyway population) currently winter in California's Central Valley (Baldassarre, 2014).

Farther south, lettuce and alfalfa fields attract large numbers of wigeons to California's Imperial Valley, and good numbers winter along the coast and interior of western Mexico, especially where pondweeds are abundant (Leopold, 1959).

In the Central Flyway, the playas of northwestern Texas and the Gulf coast of Texas account for nearly all of the wintering birds. In the Mississippi Flyway, coastal Louisiana represents the major wintering area for wigeon, especially since recent hurricanes have created openings in the dense coastal marshlands (Baldassarre, 2014). Along the Atlantic coast wigeons winter on fresh and brackish areas from Long Island southward, particularly in Maryland, coastal North Carolina, and central Florida.

In the Chesapeake Bay region the wigeon is an abundant migrant and common winter resident (Stewart, 1962). It is most often found on fresh or brackish estuarine bays where submerged plants such as wild celery (*Vallisneria*), naiad (*Najas*), pondweeds (*Potamogeton*), and wigeon grass (*Ruppia*) are plentiful. In more salty water the birds occur where eelgrass (*Zostera*) and wigeon grass are abundant, and in marsh habitats they prefer areas containing wigeon grass or muskgrass (*Chara*).

General Biology

Age at maturity. Wigeons presumably normally nest in their first year of life. Ferguson (1966) indicated that 14 of 22 respondents to a survey indicated first-year nesting by captive birds, while six and two reported second- and third-year nesting, respectively.

Pair-bond pattern. New pair-bonds are apparently established each year; there appears to be no evidence of pairs remaining intact between breeding seasons. In the closely related Chiloe wigeon (*Anas sibilatrix*) of South America, the male regularly participates in brood care and presumably has a more persistent pair-bond; male American wigeons have also occasionally been seen accompanying broods (Wishart, 1983a).

Fig. 9. Male American wigeon landing.

Nest location. Girard (1941) noted that among 45 nests the average distance from water was 98 yards and the range was 2 to 350 yards. Keith (1961) noted an average distance of 72 feet and an average relative light penetration of 47 percent at the floor of the nest. He found that 81 percent of 21 nests were in *Juncus* cover, while the rest were in mixed prairie or weeds. Munro (1949) also stated that nests are frequently placed in sedge meadows, and Phillips (1924) mentioned that the nest is often located at the base of a tree. Wishart (1983a) noted that 80 percent of the 30 nests he studied in Saskatchewan were under dense stands of shrubs. Wigeons also often nest on islands having colonies of California gulls, ring-billed gulls, or common terns and possibly gain some protection from predators through the proximity of these highly protective birds, although most gulls are serious egg predators.

Clutch size. Girard (1941) reported an average clutch size of 9.55 eggs for 45 nests in Montana. Keith (1961) noted an average of 8.9 eggs for 20 Alberta nests. In one study (Wishart, 1983a), first-year females laid smaller clutches (seven nests averaged 8.4 eggs) than did older hens (six nests averaged 9.8 eggs). Almost no data are available on renesting incidence or the clutch sizes of such renests; wigeon nests are so difficult to find that information on renesting is very limited, but it is known to occur, at least when nests are disturbed early in incubation (Smith, 1971).

Incubation period. Hochbaum (1944) reported a 23-day incubation period, based on a single clutch. Scott and Boyd (1957) reported a 22- to 25-day range for eggs from captive birds, probably under broody hens. Johnstone (1970) noted a 24-day incubation period for eggs of captive females.

Fledging period. Lee (1964) estimated a 47- to 50-day fledging period. Fledging in southern Canada and northern plains region requires 45 to 58 days versus 37 to 48 days in Alaska (Mowbray, 1999), where longer periods of daylight allow for much extended daily foraging opportunities.

Nest and egg losses. Estimated nest success from various regions include 20 percent for 20 Alberta nests, 38 percent for 18 Alberta nests, 43 percent for Saskatchewan nests, 57 percent for Alaska nests, 58 percent for 20 North Dakota nests, and 75 percent for 45 Montana nests (Mowbray, 1999). An average of 7.15 eggs hatched in the successful nests Girard (1941) studied, which represented 75 percent of the 45 total nests he found, with crows and skunks known nest predators.

Juvenile mortality. Keith (1961) noted that the average brood size of 75 Class I (early downy stage) broods he saw was 7.2 young, representing a 19 percent reduction from the average clutch size (8.9 eggs) that he had observed. Lee et al. (1965) reported an average brood size of 7.6 among 106 broods of all ages. Yocom (1951) noted that 13 broods ranging from nearly grown to fully grown averaged 5.6 young. More than 280 broods from various regions that were from 42 to 50 days old averaged 6.5 young (Bellrose, 1980). However, brood amalgamation is known to sometimes occur (Bellrose, 1980; Campbell et al., 1990) and can unduly influence resulting estimates of brood survival.

Adult mortality. Reinecker (1976) estimated survival rates from banding data from more than 32,000 wintering wigeons in California, reporting that mean annual survival was 66 percent for adult males and 58 percent for adult females, and their respective estimated mean lifespans were 2.3 and 1.7 years. Using resighting data from females marked with nasal collars, Arnold and Clark (1996) calculated mean annual survival rates of 6.5 percent for adult females and 57.1 percent for juvenile females.

General Ecology

Food and foraging. To a much greater extent than any other North American surface-feeding duck, the American wigeon is a grazing bird and one dependent on the vegetative parts of aquatic plants. Animal materials play a very small role in adult food consumption, although they are the prime food of ducklings (Munro, 1949). In most areas, wigeon grass and pondweed seeds and vegetative parts are prime components

American wigeon, adult male wing-flapping

of the wigeon's diet (Martin, Zim, and Nelson, 1951), supplemented by a large variety of other, mostly freshwater, aquatic plants. Cultivated crops such as lettuce, alfalfa, barley, and others are sometimes utilized heavily on wintering areas where they are readily available.

The tendency of American wigeon to feed on the aquatic plants brought up by diving ducks such as canvasbacks has long been recognized, and the ecological distribution of these two species on their migration routes and wintering grounds is quite similar (Stewart, 1962). Stewart noted that virtually all of more than 150 digestive tract contents he examined contained leaves, stems, and rootstalks of submerged plants, regardless of the habitats in which the birds were collected. Since wigeon are not among the species of surface-feeding ducks known to dive for food (Kear and Johnsgard, 1968), such underwater plants must either be reached by tipping-up or by obtaining those brought to the surface by coots, diving ducks, or swans.

Sociality, densities, territoriality. American wigeon do not usually congregate in extremely large flocks, although rich sources of foods such as lettuce fields or similar truck crops may result in fairly large numbers of birds. Jahn and Hunt (1964) noted a maximum fall concentration of 67,000 birds on Horicon National Wildlife Refuge and noted the birds' attraction to large open-water lakes with extensive beds of submerged plants.

During spring migration, wigeon usually move north in small groups. Munro (1949) mentioned that spring flocks often numbered ten or fewer birds. Wigeon often mingle with gadwalls at this time, as well as with coots and diving ducks, and they may try to steal and eat green vegetation that the coots and diving ducks bring up from below.

Little information is available on breeding densities. Keith (1961) reported a five-year average of 5 wigeon pairs using a 680-acre study area in Alberta, or almost 5 pairs per square mile. If only water acreage is considered, this density would represent about 3.6 pairs per 100 acres. A maximum brood density of 0.45 broods per acre has been reported for a 20-acre marsh in northern Michigan (Beard, 1964). Estimates of home ranges and territory sizes are apparently not yet available.

Interspecific relationships. Because of its relatively unique foraging adaptations, there is probably little if any food competition between wigeon and other surface-feeding ducks, and certainly the availability of nest sites is not a limiting factor for wigeon. The wigeon's most important relationships with other waterfowl are with canvasbacks, redheads, whistling swans, and coots, all of which bring to the surface submerged plant materials. The ability of the wigeon to steal such materials from other birds has earned it the name "poacher."

Perhaps because their nests are usually so well hidden, wigeon are little affected by social parasitism or parasitic egg-laying by other species. Weller (1959) noted only two cases (involving the shoveler and the white-winged scoter) of other species depositing their eggs in wigeon nests, although the lesser scaup has also been reportedly implicated (Palmer 1976).

Predators of eggs and young are probably much the same as for other surface-feeding ducks, but too few wigeon nests have been studied for definite statements on this point. Evidently crows and skunks do take some eggs (Girard, 1941).

General activity patterns and movements. Few specific data are available on the daily activity rhythms of wigeon. During fall migration, there appears to be a differential sex movement. Male wigeon leave the Delta, Manitoba, region shortly after completing their molt, and early arrivals in Wisconsin are mostly adult males (Jahn and Hunt, 1864). On the other hand, concentrations of immature males and females have been found in other areas, suggesting possible different fall migration routes.

Spring counts in Washington (Johnsgard and Buss, 1956) indicated that early migrants had more nearly equal sex ratios than did later ones, suggesting that paired birds move north faster than the unpaired. Likewise, Beer (1945) observed that paired wigeon were the first to depart from their wintering grounds in southwestern Washington.

Social and Sexual Behavior

Flocking behavior. Wigeons are not highly social; flock sizes during spring migration are usually rather small. In southern Canada breeding areas, the birds are seen in small flocks that move widely over the breeding area, with average home ranges of about 60 acres (25.3 hectares). By mid-May pairs begin to separate out from the flocks and establish breeding territories that average about 20 acres (7.8 hectares) (Wishart, 1983a).

American wigeon, adult male

Pair-forming behavior. Most pairing occurs on the wintering grounds, prior to the start of northward migration. However, there is probably some separation of pair members, and the remaining unpaired males continue to vie for the available females throughout the migration period. Aquatic courtship is marked by ritualized aggression in the form of gaping and raising of the folded wings, and an important aspect of pair formation is the combination of inciting by females and turning-of-the-back-of-the-head by males (Johnsgard, 1960, 1965). Inciting may also occur during aerial chases; Hochbaum (1944) mentioned wigeon hens reaching back laterally to "bill" one of the chasing males. Many such aerial chases originate as, or develop into, attempted rape chases, and their role in pair formation is probably questionable.

Copulatory behavior. Copulation is preceded by mutual head-pumping, and in the single instance of observing a completed copulation, I noted that the male turned and faced the female while remaining in an erect posture for several seconds afterward (Johnsgard, 1965).

Nesting and brooding behavior. Incubation begins with the laying of the last egg and is undertaken by the female alone. After hatching, the female leads her young into open-water areas such as marsh-lined ponds. For the first several weeks the young are entirely surface-gleaners and dabblers, slowly and deliberately moving through the marsh. When about four weeks old, they begin to tip-up for food. Brooding female wigeon are among the noisiest of ducks, and when their brood is threatened females typically remain behind, quacking loudly while the young escape to cover. This distraction behavior may last 15 minutes or more. Only when the young are nearly grown is the female usually silent (Beard, 1964). Beard also reported that female wigeons were highly aggressive toward strange ducklings, even of their own species. Of 16 cases of young being driven away by female wigeons, 15 involved other wigeons' ducklings. If the young duckling survived the first few attacks and persisted in following the brood, it was frequently accepted.

Postbreeding behavior. Adult males leave their breeding grounds in southern Manitoba in late August and early September, and soon thereafter wigeons begin to concentrate in such northern states as Wisconsin, where they gather on areas that provide a combination of protection from disturbance and a supply of submerged aquatic foods. Apparently, in certain localities there is a differential migration of immature male and female wigeons (Jahn and Hunt, 1964).

Falcated Duck
Anas falcata Georgi 1775

Other vernacular names. Bronze-capped teal, falcated teal

Range. Breeds in Asia south of the Arctic Circle from the Upper Yenisei River (Noyarsk) east to to Manchuria, Russia's Pacific coast, Hokaido, and Sakhalin, and south to Lake Baikal and northeastern Mongolia. Winters in Japan, Korea, and eastern and southern China south to Burma (Myanmar), with occasional stragglers reaching western North America, especially Alaska.

Subspecies. None recognized.

Measurements. *Folded wing:* Delacour (1954): Males 230–242, females 225–235 mm. Kear (2005): Males 242–268 mm, average (10) 253 mm; females 226–236 mm, average 231 mm.
  Culmen (bill). Delacour (1954): Males 40–42 mm, females 38–40 mm. Kear (2005): Males 39–48 mm, average (10) 42.8 mm; females 36–40 mm, average 38.8 mm.

Weights (mass). Dementiev and Gladkov (1967): Males average ca. 750 g, females 640–660 g. Chen Tsohsin (1963): 10 males average 713 g (590–770 g); 5 females average 585 g (422–700 g). Kear (2005): 4 males 590–770 g, average 713 g; 5 females 422–700 g, average 585 g.

Identification

In the hand. Both sexes of this rare dabbling duck are similar to wigeon and also have a greenish speculum. But there is no black anterior border on the greater coverts, and the coverts are never pure white, only grayish to grayish brown. The elongated sickle-shaped tertials on the male are unique and by themselves will identify that species, but females lack these ornamental features. The brownish underparts of females, their longer bill (over 35 mm), and the presence of a rudimentary crest serve to separate them from female American wigeons.

In the field. Males in nuptial plumage—with their long, bronzy to green-glossed crest; a gadwall-like "scaly" breast pattern; and long sickle-shaped tertials that nearly reach the water in swimming birds—are distinctive. The species is so rare in North America that lone females should not be identified in the field because they closely resemble female wigeon and gadwalls.

Fig. 10. Male falcated duck.

Age and Sex Criteria

Sex determination. Except when in eclipse (basic) plumage, the presence of sickle-shaped tertial feathers serve to distinguish adult males from females. In eclipse plumage, a brighter speculum pattern and a slight iridescence on the head may identify males.

Age determination. Immature males resemble females but have a dark crown with a greenish cast (Kear, 2005). No doubt the notched tail criterion will serve to identify immature birds of both sexes through much of their first fall of life.

Identification

In the hand. *Males* in breeding plumage have a strongly crested head that is mostly iridescent bronzy green and chestnut purple tints but with a white spot above the base of the upper mandible. The throat and foreneck are also white, with a narrow green collar. The body plumage is primarily gray, with fine black

Falcated duck, adult male

vermiculations and with the black on the breast forming crescents. The under tail-coverts are patterned with two buff triangles, separated medially and anteriorly with black, while the tail is gray, edged with white. The upper tail-coverts are gray and black, the coverts very long and partially hiding the tail. The longer scapulars and tertials are gray and black, the latter being greatly extended and curved down over the other wing feathers. The speculum is iridescent green, bounded in front and behind with white lines, while the upper wing-coverts are gray. The iris is brown, the bill blackish, and the legs and feet are bluish gray to yellowish. *Males in eclipse plumage* resemble females but have a more brilliant speculum and grayish rather than brownish upper wing-coverts. *Females* are mostly brown and gadwall-like, but a small crest is present and the speculum is iridescent green, as in males. The mandible is spotted with black, and yellow is extensive on the lower mandible. *Juveniles* resemble adult females but lack the nape crest.

In the field. Falcated ducks are best recognized by the distinctive shape and plumage of the male, but females are gadwall-like. The females might also be mistaken for the American wigeon, but their longer bills and rudimentary crests should help to separate them. The calls of the female are generally like those of a gadwall; the decrescendo call is from 2 to 5 syllables in length. Males utter a high-pitched whistle, *lililili*, and also produce a vibrating *rruh-urr* call during display.

Occurrence in North America

Like the Baikal teal, most records of this Asian species have come from Alaska. Gabrielson and Lincoln (1959) listed two of these; a male that was collected on St. George Island and a pair seen at Attu Island. Two males were later collected, and several more were seen at Adak Island (Byrd et al., 1974).

The first Canadian record was of a male that was observed near Vernon, British Columbia, in 1932 (Godfrey, 1966). Other early records include a sighting from San Francisco, California, and one from Roaches Run, Virginia. By 2007 there had been multiple records for British Columbia, and at least three records for Washington state.

A recent (2016) eBird map indicated several dozen sightings, with several locations in California, including Colusa National Wildlife Refuge, Honey Lake Wildlife Area, and Upper Newport Beach Nature Preserve. There are also sight records from Fern Ridge Wildlife Management Area, Oregon; Samish Flats, Washington; Tofino, British Columbia; and Attu Island, Alaska.

Gadwall
Anas strepera Linnaeus 1758

Other vernacular names. Gray duck

Range. Breeds throughout much of the Northern Hemisphere, in North America from Alaska south to California and from Quebec south to North Carolina; also breeds in Iceland, the British Isles, Europe, and across temperate parts of Asia to eastern China and Japan. Winters in North America from coastal Alaska south to southern Mexico, the Gulf coast, and along the Atlantic coast to southern New England.

North American subspecies. *Anas s. strepera* L.: Common gadwall. Range as indicated above.

Measurements. *Folded wing:* Delacour (1956): Males 260–282 mm, females 235–260 mm. Owen (1977): Adult males average 269.5 mm, females average 253.8 mm.

Culmen (bill): Delacour (1956): Males 38–45 mm, females 36–42 mm. Owen (1977): Adult males average 43.1 mm, females average 42.5 mm.

Weights (mass). Bellrose and Hawkins (1947): 16 adult males, average 2.18 lb. (989 g); 68 immature males, average 2 lb. (907 g); 14 adult females, average 1.87 lb. (848 g); 66 immature females, average 1.78 lb. (807 g). Nelson and Martin (1953): 104 males, average 2 lb. (906 g); 89 females, average 1.8 lb. (815 g). Owen (1977): Adult males average 766 g, females average 699 g.

Identification

In the hand. Positive identification of gadwalls in the hand is simple: they are the only dabbling ducks with several secondaries entirely white on the exposed webs, the remaining secondaries being black or grayish. Confirming criteria are the yellow legs and slate gray (males) or gray and yellowish (females) bill color, a white abdomen, and the usual presence of some chestnut coloration on the upper wing-coverts.

In the field. Although one of the easiest species of ducks to identify in the hand, gadwalls are perhaps the American dabbling duck that is most commonly misidentified or unidentified because of the species' lack of brilliant coloration. *Breeding males* appear to have an almost entirely gray body, except for the black hindquarters, which are apparent at great distances. In spring, the upper half of the head appears to be a considerably darker shade of brown than the lower part of the head and neck, but during fall this difference is not so apparent.

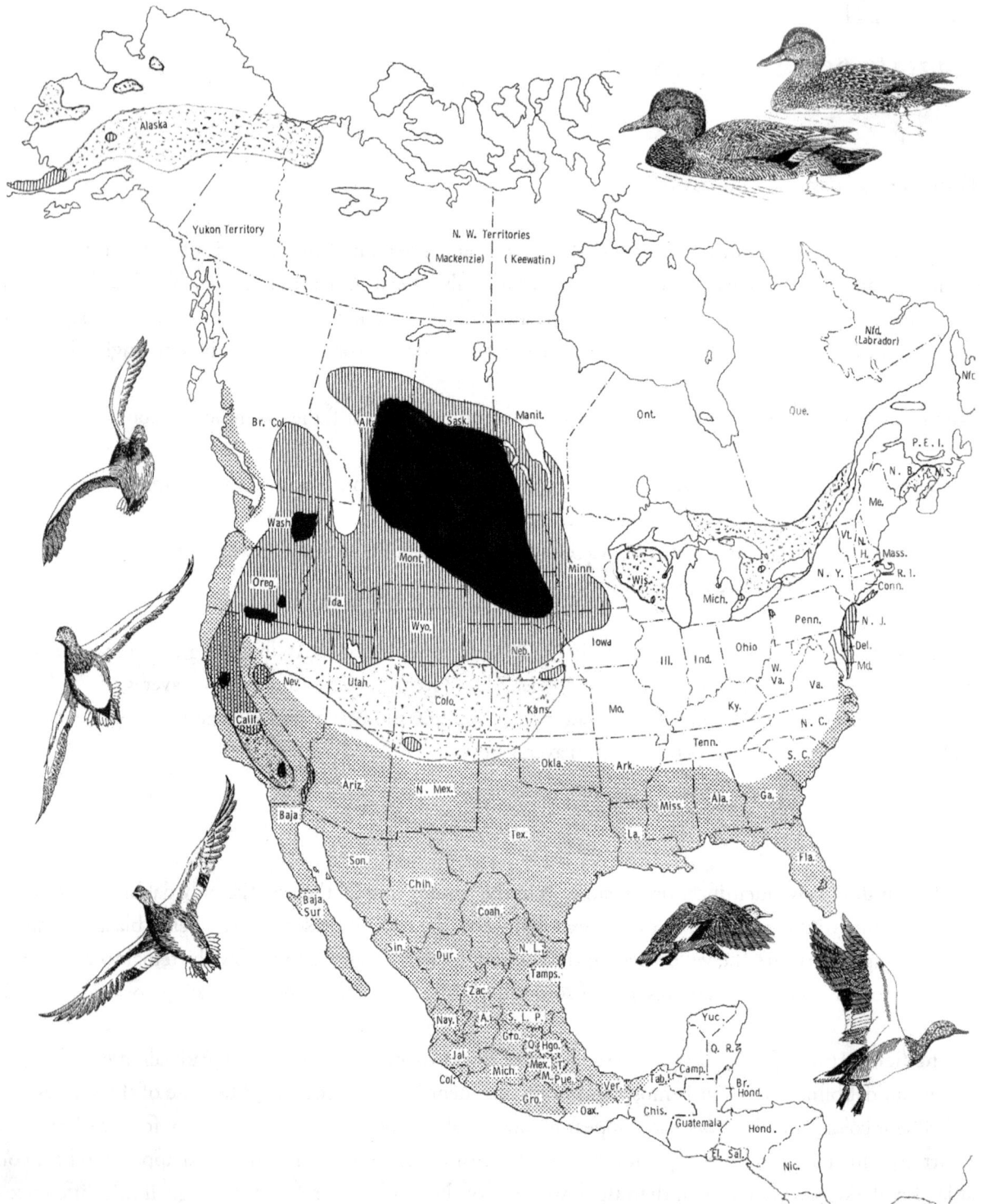

The breeding (vertical hatching, with denser concentrations inked), wintering (shaded), and acquired or marginal (stippled) range of the gadwall.

Females are best recognized by their association with a male, but at fairly close range the yellowish sides of her otherwise gray bill can be seen, and the bill is clearly shorter and weaker than that of a female mallard, which she closely resembles. The white secondaries are usually not visible when the birds are at rest. However, the white secondary pattern is highly conspicuous during flight, with white also appearing on the underparts of the body and on the under wing-coverts, the rest of the bird appearing brownish. From early fall until spring, the courting calls of the males can be heard, either when in flight or on the water, a combination of low-pitched *raeb* notes interspersed with *zee* whistles, often in a distinctive *raeb-zee-zee-raeb-raeb* cadence (on the water only). The female has various mallard-like quacking notes, including a series of paced *quack* notes when alarmed, or a decrescendo series of notes that are somewhat more rapid and higher pitched than occurs in mallards.

Age and Sex Criteria

Sex determination. *Adult males* have some black or chestnut middle coverts that are not pale-edged, while females have only a few black or chestnut coverts that are limited to a few rows and are pale-edged or barred. The tertials of adult males are long, pointed, and silver gray. The presence of vermiculations—narrow, dark, wavy, and "worm-like" patterning—anywhere on the body indicates a male, as do chestnut-tipped longer scapulars. The greater tertial coverts of adult males rarely have any white tipping, while those of females do. *Adult females* have shorter, more bluntly pointed tertials that are silver-brown with cream-colored tips. *Immature females* usually lack chestnut on their middle coverts, but immature males usually have some chestnut present (Carney, 1964). Juveniles of both sexes may have short, bluntly pointed, and frayed tertials.

Age determination. In *immatures* the juvenal tertials of both sexes are short, bluntly pointed, and usually frayed at their tips. The juvenal greater tertial coverts of both sexes are partly black and partly gray and, as in adult females, are usually tipped with white. However, immature males have narrower coverts with less white tipping than those of adult females (Carney, 1964). The tips of some tail feathers may be notched in immatures; Oring (1968) reported that these feathers are lost in an asymmetric fashion between September and February, but females may retain some juvenal tail feathers until spring.

Distribution and Habitat

Breeding distribution and habitat. The gadwall is distinctly westerly and southerly in its primary North American breeding distribution, with only scattered nesting records from Alaska. These are mostly from the Alaska Peninsula and along the southern coast east to somewhat beyond the Copper River (Gabrielson and Lincoln, 1959; Kessel and Gibson, 1978; Leschack, McKnight, and Hepp, 1997). Canadian breeding is largely limited to southern and southeastern British Columbia and the grassland areas of Alberta, Saskatchewan, and Manitoba, with local breeding extending eastward across southern Ontario and southern and southeastern Quebec (St. Lawrence River valley) to the Maritime Provinces (Godfrey, 1986).

In the United States, breeding extends from Washington south to southern California and eastward through northern parts of Nevada, Arizona, New Mexico, and Texas, then northeast through Kansas, Nebraska, the Dakotas, and western Minnesota, with scattered breeding to the east in Wisconsin, Michigan, and New York (Leschack, McKnight, and Hepp, 1997; Baldassarre, 2014).

Beginning in the mid-1940s, gadwalls began nesting at Jones Beach, Long Island, and substantial populations soon developed there (Sedwitz, 1958). Additional breeding populations developed at Pea Island National Wildlife Refuge, North Carolina, and later at Brigantine National Wildlife Refuge, New Jersey. There are also breeding records from Massachusetts, Connecticut, Pennsylvania, New Jersey, Delaware, Maryland, and Virginia, mostly in coastal grassland habitat. Some of these nestings may have resulted from the introduction of gadwalls in New England (Borden and Hochbaum, 1966). Henny and Holgersen (1974) documented this eastern range expansion.

The breeding habitat of gadwalls is typically marshes or small lakes in grassland. In particular, the presence of grassy islands is of considerable significance in determining nest distribution and density. Alkaline marshes seem to be preferred over those with low salt concentrations. Drewien and Springer (1969) noted that during two years of study, breeding gadwalls were consistently more numerous on shallow prairie marshes than on temporary water areas, or shallow to deep marshes, or deep and open-water marshes. Preferred nesting cover consists of dense, coarse vegetation, and the presence of herbaceous weeds interspersed with shorter vegetation on islands surrounded by open water may facilitate colonial nesting (Duebbert, 1966). Heavy grass or brush, such as might be provided by shrubby willows, is also an important nesting cover.

Population. North American breeding grounds surveys in 2014 indicated a total population of 3.8 million birds, 102 percent above the long-term average of 1.9 million (USFWS, 2014). The estimated US hunter-kill in 2014 was about 1.58 million birds and in Canada was about 39,000 birds.

Wintering distribution and habitat. Wintering occurs over much of the United States from Prince William Sound and Kodiak Island south along coastal British Columbia to southern Mexico, where it is abundant on both coasts and the interior, with historically large concentrations on the Nayarit coast (Leopold, 1959), but far smaller numbers have been present in recent years. Winter surveys up to the mid-1970s averaged about 24,000 birds on the east coast and somewhat fewer on the west coast; about 70,000 were estimated for the entire country from 1981 to 2000 (Baldassarre, 2014).

By flyways, the largest concentration of wintering gadwalls in the United States had long occurred in the Mississippi Flyway, in spite of most North American breeding occuring within the limits of the Central Flyway. Between 2000 and 2010, nearly half of the gadwalls counted during all Midwinter Surveys were supported by the Mississippi Flyway, a third in the Central Flyway, nearly 10 percent in the Pacific Flyway, and less than 2 percent in the Atlantic Flyway. Louisiana supported over 80 percent of the birds wintering in the Mississippi Flyway (Baldassarre, 2014), mostly in coastal marshes. Many of the Central Flyway gadwalls also winter on the Texas coast (which state supported about a third of the total survey estimate), or continue on into Mexico.

Fig. 11. Male gadwall.

In the Chesapeake Bay area, migrant and wintering gadwalls usually are found on slightly brackish estuarine bays, where there are such submerged plants as wigeon grass (*Ruppia*), clasping-leaf pondweed (*Potamogeton perfoliatus*), and water milfoil (*Myriophyllum*). They also occur on natural ponds and marsh impoundments where wigeon grass and muskgrass (*Chara*) are the most common submerged plants (Stewart, 1962).

General Biology

Age at maturity. Of 22 responses to a survey of aviculturists, 13 reported breeding by captive gadwalls at one year of age, with seven and two responses indicating second- and third-year breeding, respectively (Ferguson, 1966).

Pair-bond pattern. Gadwalls renew their pair-bonds every year, and these are terminated early in the incubation period, when males desert their mates and begin their postnuptial molt.

Nest location. In a study involving 660 nests, Williams and Marshall (1938) noted that the three most preferred cover types, in sequence of decreasing importance, were hardstem bulrush (*Scirpus acutus*), brushy willows (*Salix* spp.), and various herbaceous weeds. In a sample of 381 nests studied by Miller and Collins (1954), nettle was a highly preferred nest site. They found that about 84 percent of the nests were in vegetative cover between 13 and 36 inches high, and about 44 percent of the nests were on islands. Nest concealment was very high, with more than 90 percent of the nests concealed on all four sides and about 70 percent concealed from above as well. About 85 percent were located from 3 to 50 yards from water.

In some studies (Keith, 1961; Hunt and Naylor, 1955), the use of weeds as nest cover is of equal or lesser importance than that provided by Baltic rush (*Juncus balticus*), which often occurs as a shoreline belt around prairie marshes. However, Gates's (1962) study in Utah showed a clear preference by females for dry over wet sites and upland vegetation over lowland cover types, with the densest and driest cover types generally being selected.

Clutch size. A variety of clutch size samples from North American gadwalls indicate an average between 9 and 11 eggs. Miller and Collins (1954) reported an average clutch of 11.0 eggs in 344 nests, similar to Gates's (1962) estimate of 11.1 eggs in 141 early nests. Similarly, Sowls (1955) noted an average clutch of 10.5 eggs for 17 early nests. Keith (1961) noted a decrease in clutch size from about 10 to 9 eggs as the breeding season progressed in Alberta. Williams and Marshall (1938) indicated a modal clutch size of 10 eggs in Utah, but the average of 660 nests was 9.09 eggs, probably reflecting renesting influences.

The renesting incidence has been estimated to be 82 percent in Alberta and 96 percent in Utah. Duebbert (1966) noted an average clutch of 9.6 for 140 clutches in a colonial nesting situation in North Dakota, but his indicated clutch range of 5 to 20 eggs and comment on egg variability suggest that parasitic egg-laying probably influenced his data. Eggs are laid at a daily rate (Gates, 1962). Gates reported that renests, up to three of which were found, averaged 7.8 eggs, as compared with 10.7 eggs in initial nesting attempts by the same birds.

Incubation period. Normally there is a 26-day incubation period, although there are records of 25-day and 27-day incubation periods (Bauer and Glutz, 1968) as well as a case of a nest hatching after 29 days, during which incubation was abnormally disturbed (Duebbert, 1966). Vermeer (1968) calculated an average period of 25.1 days based on a sample of ten clutches, with a range of 22 to 27 days. Oring (1969) reported a 24-day average period for incubator-hatched gadwalls and a 25.7-day average for clutches hatched under natural conditions.

Fledging period. Hochbaum (1944) estimated a fledging period of 49 to 63 days. Most other published estimates are for seven weeks. Oring (1968) reported the first flight in 47 of 50 hand-reared gadwalls at between 50 and 56 days of age.

Nest and egg losses. Nesting success no doubt varies greatly with time and locality, but some high nesting success rates have been reported. Duebbert (1966) noted a nesting success averaging nearly 90 percent during two years for an island-nesting population. A similar 90 percent nesting success was noted for 381

Gadwall, adult pair

nests in California (Miller and Collins, 1954). In Utah, Williams and Marshall (1938) reported an 85 percent hatching success for a sample of 6,000 eggs. Keith (1961) found a much lower nest success in Alberta but estimated that, with renesting included, 45 percent of the females in his study area eventually successfully hatched a brood. Vermeer (1970) reported a nest success of only 33.3 percent for one group of island-nesting gadwalls in Alberta, as compared with an earlier (1968) nesting success of 90.0 on a different island. He found (1970) that gadwalls nested in higher densities in the presence of terns (*Sterna*) and probably also gulls (*Larus*), although some species of gulls may cause heavy egg and chick mortality. Oring (1969) found an overall nesting success of 46 percent for 30 nests, with losses to ground squirrels, raccoons, and skunks.

Juvenile mortality. Fledging success from 26 gadwall broods studied by Vermeer (1968) was nil because of high predation on ducklings by California gulls (*L. californicus*). Gates (1962) estimated an average pre-fledging duckling mortality of 23 percent, with most losses occurring in the first 18 days of life. The most important duckling predators in this area were also California gulls. Gates (1962) calculated a first-year mortality rate of 67 percent for birds banded as juveniles.

Adult mortality. Gates (1962) estimated an annual adult mortality rate of 52 percent for birds banded as adults or not aged. This is identical to results obtained from banded gadwalls in England (Wainwright, 1967).

General Ecology

Food and foraging. In the several studies that have been done on gadwall foods, there has been a consistently low percentage of animal materials present and a high incidence of the vegetative parts of submerged plants. Martin et al. (1951) reported high use of wigeon grass, algae such as muskgrass (*Chara*), pondweeds, and other aquatic plants. In a fall sample of nearly 200 stomachs from Utah, Gates (1957) noted that the foods found were mostly the vegetative parts of wigeon grass, pondweeds, and horned pondweeds (*Zannichellia*) and the seeds of hardstem bulrush (*Scirpus acutus*) and salt grass (*Distichlis*). Stewart (1962) found that among gadwalls shot in brackish and freshwater estuaries of the Chesapeake Bay, vegetative parts of plants such as wigeon grass, muskgrass, eelgrass (*Zostera*), pondweeds, and naiad (*Najas*) were the principal foods.

Gadwalls are almost exclusively surface-feeders, although they have been observed diving for food on a few occasions (Kear and Johnsgard, 1968). Thus, they are largely dependent on food that they can reach by tipping-up and tend to feed in rather shallow marshes with abundant submerged plant life growing close to the surface.

Sociality, densities, territoriality. Gadwalls are relatively social on the nesting grounds, at least in island-nesting situations. Gates (1962), studying in an area where no island nesting was possible, noted a definite spacing-out of pairs and moderately large home ranges (average of five pairs per 67 acres). These home ranges overlapped considerably and shared common areas for foraging or loafing, although not simultaneously. Established males attempted to discourage new pairs from breeding in the same area but were often unsuccessful.

In island-nesting situations, territoriality is essentially absent, and Duebbert (1966) believed that gadwalls have evolved behavior patterns that enable many pairs to nest in a very restricted area. He noted nest densities of 78 and 121 nests on a seven-acre island at Lower Souris National Wildlife Refuge, North Dakota, during two summers, and there is an earlier record of 106 nests on a one-half-acre island.

Interspecific relationships. Nest site competition with other ducks is probably not significant for gadwalls. Gates (1962) noted that other species of surface-feeding ducks seem less dependent than the gadwall on dry and/or dense cover for nesting. Vermeer (1968) noted a fairly low rate of nest parasitism in gadwall nests, with 11 of 54 nests being affected. These were mostly by lesser scaup and white-winged scoters. There is also a reported case of nest parasitism by redheads (Weller, 1959).

Gadwalls have been shown to exhibit a preference for nesting among tern colonies (Vermeer, 1970) in Alberta. Likewise, in Russia, nesting has been noticed among gulls, terns, plovers, and other shorebirds (Dementiev and Gladkov, 1967). Bengtson (1972) found that predation by common ravens was the greatest single cause of nesting failure, while mink, parasitic jaegers, and great black-backed gulls also accounted for some losses. Egg predation by California gulls on gadwall nests is sometimes extremely high, and they may also be responsible for duckling losses (Odin, 1957).

General activity patterns and movements. In contrast to mallards and pintails, gadwalls typically exhibit a considerable delay between the arrival at their nesting grounds and the beginning of nesting. Gates (1962) noted an average post-arrival period of 17 days prior to establishment on a breeding home range, and another prenesting period of 11 days before the beginning of egg-laying. This delay is apparently related to the gadwall's choice of dry and dense nesting cover. During this prenesting period, paired birds remain gregarious until the home range area is established, and pairs may forage and loaf together. Gates found that home ranges of gadwalls in Utah ranged from 34 to 87 acres, with nests well scattered, whereas Duebbert (1966) found that much larger home ranges occurred among a group of colony-nesting gadwalls. There, incubating females sometimes flew more than a mile to rest and feed unmolested by non-pair drakes.

Social and Sexual Behavior

Flocking behavior. Although gadwalls nest relatively late, an early reestablishment of bisexual flocks in fall is typical. This seems to be related to the fact that gadwalls begin pair formation activities unusually early, even while males are still in eclipse plumage. In Austria this activity begins in August, and within a month 50 percent to 70 percent of the females appear to be paired. Thus, the fall migration of this species does not show sexual segregation, at least by comparison with many other surface-feeding ducks (Bezzel, 1959). Sex-ratio counts made during early and later stages of migration also do not show changes suggestive of differential sex migration or earlier migration of paired birds (Johnsgard and Buss, 1956).

Pair-forming behavior. Although pair-forming behavior in gadwalls begins unusually early, and most aquatic courtship has occurred prior to the acquisition of the males' nuptial plumage, there is a secondary spring peak of social courtship (Bezzel, 1959). Aerial chases also progressively increase toward spring, with a peak (in Austria) in May, or just prior to the onset of incubation. Duebbert (1966) also noted a high intensity of aerial chases in North Dakota in late May and early June, when many paired birds moved to the nesting island and egg-laying began. Flights continued throughout most of the incubation period of July. Duebbert interpreted the earlier pursuit flights as a reflection of individual intolerance, and the later ones as increasingly sexual. It is unlikely that such aerial chases play any functional role in normal pair formation, but rather pairs seem to be formed by the combination of female inciting and male turning-of-the-back-of-the-head displays, as in other ducks that have been studied (Johnsgard, 1965, see Figs. 6 and 12).

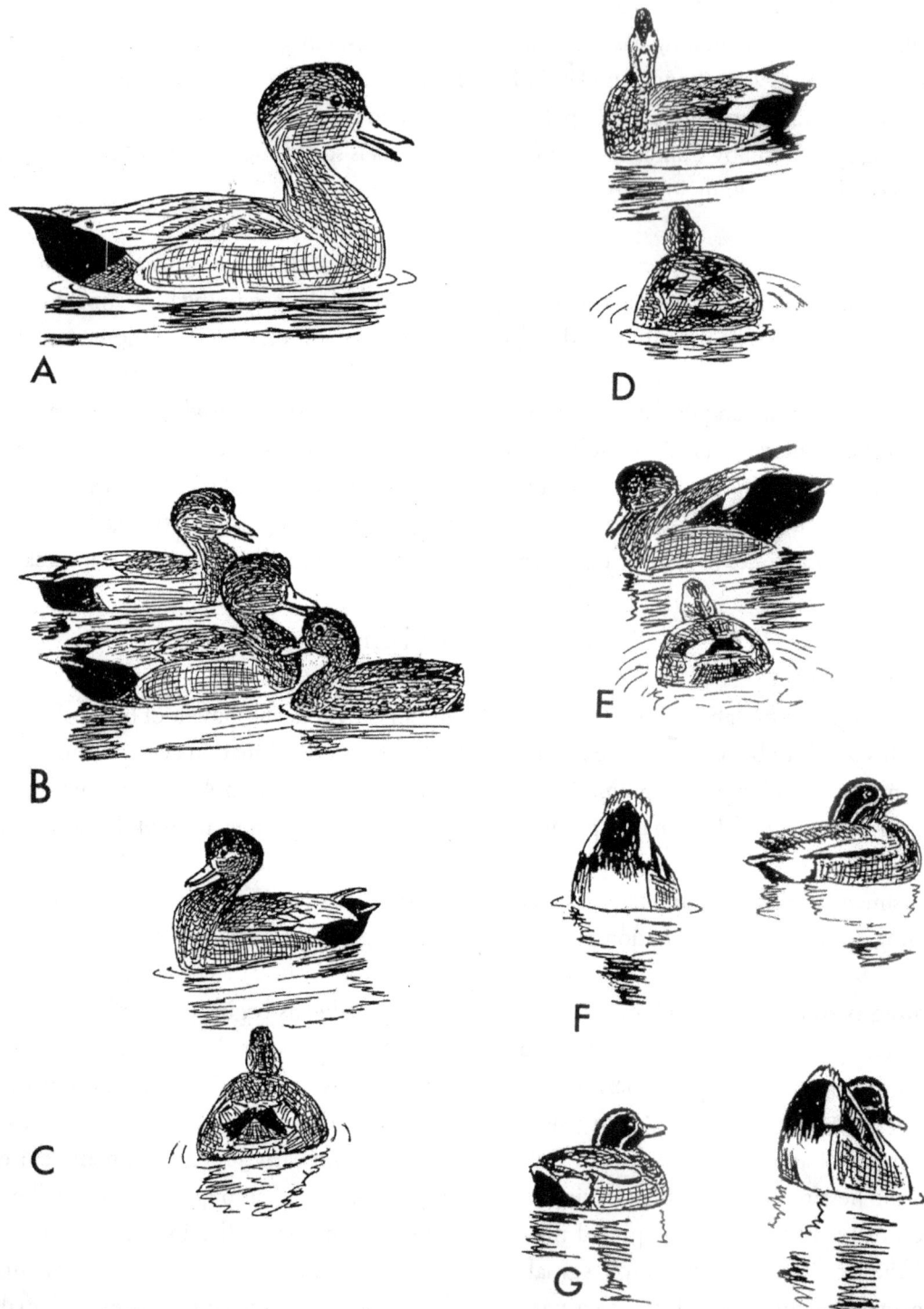

Fig. 12. Sexual behavior of gadwall (A–E) and green-winged teal (F–G), including (A and B) burping, (C–D), head-up-tail-up, (E) down-up, and (F–G) down-up.

Fig. 13. Three-bird courting flight, gadwall.

Copulatory behavior. As in other surface-feeding ducks, copulation is preceded by mutual head-pumping behavior. Following treading, the male utters a whistle-grunt call and then turns and faces the female in a motionless and erect posture (Johnsgard, 1965).

Nesting and brooding behavior. When looking for nest sites, the pair may fly out to grassy areas and land together. While the male waits, the female walks into the weedy growth. This phase may precede actual egg-laying by 5 to 7 days (Duebbert, 1966). When laying, females go to their nest sites between 5:00 a.m. and 7:00 a.m., either flying to a point up to 25 feet away and walking the remaining distance to the nest or, in the case of nests in tall cover, landing within a few inches of the nest's location. Duebbert found that the male may desert the female as early as the seventh day of egg-laying, or he might remain until the day prior to the hatching of the eggs. Gates (1962) indicated that the desertion usually occurred before the midpoint of incubation.

Following hatching, females with broods move to deep-water marshes and edges of large impoundments, sometimes traveling in excess of a mile, and in one study averaging about half a mile (Gates, 1962). Gates found no evidence of brood mergers in the broods of marked hens that he studied. Shortly after or even before leaving their mates, males begin to molt. Gates found that such males retained some sexual interest, nevertheless, and that some even participated in attempted rapes of other nesting females.

Postbreeding behavior. Oring (1969) confirmed Gates's observations as to the variations in times at which males deserted their mates and believed that the sight of postbreeding groups might hasten the breakup of pairs. He also believed that some yearling males never participate in courtship display and are the first to undergo postnuptial molt of their flight feathers. They are then followed in sequence by early breeding males, later breeding males, sexually active but nonbreeding drakes, early breeding females, and finally late breeding females. Some late breeding hens may migrate to their winter quarters before undergoing their flightless period.

In Oring's study maximum molting congregations of males occurred at the end of June, and by early August about half of the adult males were flightless. At this time, captive males were not yet flightless but were exhibiting dawn and dusk periods of nervousness that seemed to be indicative of premigratory restlessness. As the wild birds regained their powers of flight they formed large, wary flocks, which foraged during the entire day if undisturbed. Most of them had left the breeding area by the end of September.

Baikal Teal
Anas formosa Georgi 1775

Other vernacular names. Clucking teal, Formosa teal, spectacled teal

Range. Breeds in eastern Siberia and northern Ussuriland, possibly also in Kamchatka. Some summer records from St. Lawrence Island, King Island, and mainland Alaska, but no records of North American breeding. Winters mainly in central China, with smaller numbers in Japan, Taiwan, and southeast Asia rarely as far as India, with rare stragglers along the Pacific coast of North America from British Columbia to California. The species has been in marked population decline since the late 1900s.

Subspecies. None recognized.

Measurements. *Folded wing:* Males 200–216 mm, females 190–198 mm. Kear (2005): 12 males 203–225 mm, average of 12, 211 mm; females 402–505 mm, average of 12, 206.5 mm.

Culmen (bill): Males 35–38 mm, females 33–36 mm. Kear (2005): 12 males 38–40 mm, average of 12, 38.7 mm; females 36–38 mm, average of 12, 37 mm.

Weights (mass). Kear (2005): Males 360–520 g, average of 12, 437 g; females 402–505 g, average of 8, 431 g.

Age and Sex Determination

Sex determination. Among adult birds, females can probably be recognized by their relatively dull speculum pattern, the absence of ornamental tertials or iridescent head-patterning, and a paler throat than occurs in eclipse plumage males.

Age determination. Immatures resemble females but have a less-defined facial pattern, and browner rather than rufous tones (Madge and Burn, 1988). First-year birds probably retain notched tail feathers through their first fall of life.

Identification

In the hand. Because of its very similar speculum patterns, the Baikal teal is most readily confused with the green-winged teal, from which it can be readily separated by its longer tail (minimum 75 mm) and larger

Fig. 14. Male Baikal teal.

size (over 400 g). The male's distinctive head pattern is usually not attained until late winter, but the orna-mental chestnut-striped scapulars and tertials are present earlier. Females should be carefully compared with the female green-winged teal, which they closely resemble but differ in their definite white (rather than buffy and faintly striped) cheek spot at the base of the upper mandible, their clearer white throat with an exten-sion up the sides of the cheeks, and the dark area above the eye that interrupts the pale superciliary stripe.

In the field. The male in nuptial plumage is unmistakable at close range. The bird sits in the water with its colorful head low on the breast, its tail well out of the water, the ornamental scapulars hanging down over the flanks, and vertical white bars visible in front of the black under tail-coverts and on the sides of the breast. Its distinctive clucking call, *ruk-ruk'*, or *ruk*, is uttered only during sexual display. The quacking notes of the female are rather infrequent. In the air females resemble a green-winged teal but have brownish gray rather than mostly white under wing-coverts. Lone females should not be identified as Baikal teal except un-der extremely favorable conditions, when their distinctive facial markings noted above can be clearly seen.

Baikal teal, adult males

Occurrence in North America

Most of the records of this beautiful Asian species of duck have originated from Alaska. Gabrielson and Lincoln (1959) summarized the majority of these, which include a male collected at Wainwright, two males collected on King Island, pairs collected on St. Lawrence Island, and a pair plus a male at Wales. As of the mid-1970s there was apparently only one Canadian record, at Ladner, British Columbia. Records from south of Canada then included specimens from Washington and Oregon as well as one or more sightings in California, New Jersey, Pennsylvania, and Ohio.

A recent (2016) eBird map indicated several dozen sightings from the Aleutian Islands including Attu Island, Shermya Island, St. George and St. Paul in the Pribilof Islands, and near Prudhoe Bay, Alaska. In British Columbia there are sightings from metropolitan Vancouver and the nearby Agassiz Wetlands. Washington records are from the Kent Ponds (near Seattle) and Columbia National Wildlife Refuge. In Montana there are sightings from Maclay Flat Recreation Area near Missoula.

Green-Winged Teal
Anas crecca Linnaeus 1758

Other vernacular names: Common teal (British Isles), greenwing, northern green-winged teal, teal (Britain)

Range: Breeds throughout much of northern Europe and Asia, the Aleutian Islands, temperate North America, and Iceland. In North America, winters from southern Canada along both coasts south through the central and southern states to Mexico and Central America.

North American subspecies (recognized by Delacour, 1956): *A. c. crecca* L.: Eurasian Green-winged Teal. Breeds in temperate and subarctic Iceland, Europe, and Asia, Seen occasionally in North America during winter, especially along the Atlantic and Pacific coasts.

A. c. nimia Friedmann: Aleutian Green-winged Teal. Resident in the Aleutian Islands, from Akutan westward. Has been reported as far south as Oregon (Gilligan et al, 1994).

A. c. carolinensis Gmelin: American Green-winged Teal. Breeds in temperate and subarctic continental North America, from north-central Alaska east to New Brunswick and Nova Scotia. Sometimes (e.g., Livezey, 1991; Johnson and Sorenson, 1999) considered as a specifically distinct taxon, *A. carolinensis*.

Measurements. *Folded wing, A. c. carolinensis:* Delacour (1956): Males 179–191 mm; females 172–183 mm. Owen (1977): Adult males of *A. c. crecca* average 189.9 mm; females average 181.9 mm.

Culmen (bill), *A. c. carolinensis:* Delacour (1956): Males 34–37 mm; females 33–36 mm. Owen (1977): Adult males of *A. c. crecca* average 36.9 mm; females average 33.5 mm.

Weights (mass). Nelson and Martin (1953): Average of 199 *A. c. carolinensis* males, 0.8 lb. (362 g); average of 81 females, 0.7 lb. (317 g). Jahn and Hunt (1964): Average of 45 adult and 149 immature fall *A. c. carolinensis* males, 12 oz. (340 g); average of 33 adult and 114 immature females, 11 oz. (312 g). Owen (1977): Adult males of *A. c. crecca* average 293 g, females average 276 g.

Identification

In the hand. This species is the smallest of the North American dabbling ducks, rarely if ever exceeding a pound (453 g) in weight and with a tail of less than three inches (75 mm). The bill is relatively long (33–37 mm) but unusually narrow (12–14 mm). Besides the small size, the presence of a speculum pattern that is green inwardly, black outwardly, narrowly edged behind with white, and with a brownish anterior border is relatively diagnostic. A similar speculum pattern occurs only in the rare Baikal teal.

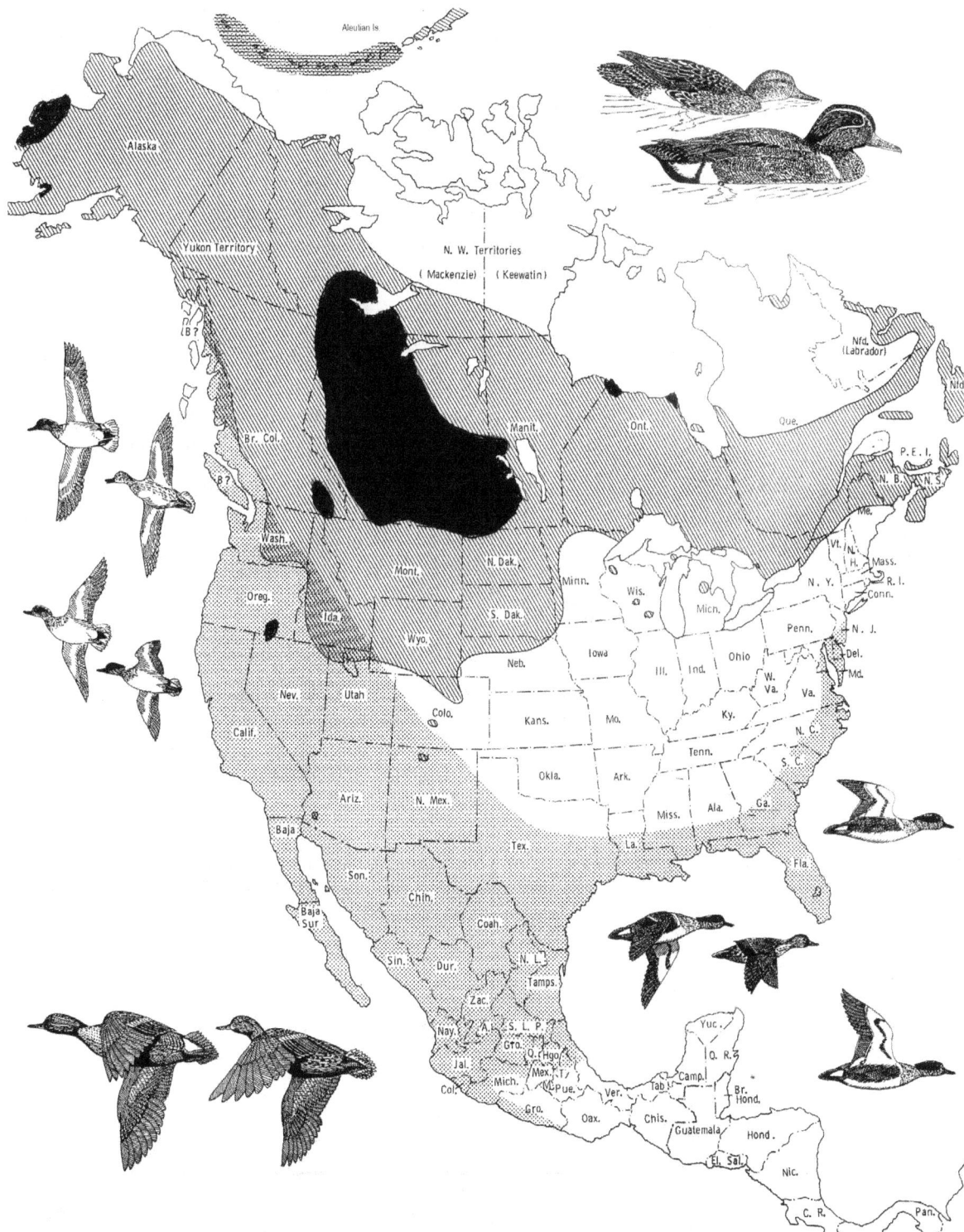

The breeding (diagonal hatching, with denser concentrations inked), wintering (shaded), and acquired or marginal (stippled) range of the green-winged teal.

In the field. Green-winged teal float lightly in the water, the tail usually well above the water, and males exhibit buffy yellow triangular patches on the black under tail-coverts. The only white marking shown by males is the vertical bar in front of the gray sides (usually) or a horizontal white stripe between the back and flanks (in the rare Eurasian and Aleutian forms). In good light, the iridescent green head patch may be distinguished from the otherwise chestnut head, the two areas separated by a narrow and often faint buffy white stripe (brighter in the Eurasian and Aleutian forms). Field recognition of the Aleutian and Eurasian forms must be based on males; females can scarcely be distinguished in the hand. In the field, lone female green-winged teal might be identified by their very small size, short neck, and overall brownish color, the head having a darker eye-stripe and a paler area near the base of the bill, but they appear much like the blue-winged teal.

In flight, green-winged teal are the essence of agility, twisting and turning like shorebirds, and alternately flashing their white under wing-coverts and dark brownish upper wing. The brown upper wing-coverts are the best way to separate green-winged teal from blue-winged or cinnamon teal in flight, although green-winged teal have shorter necks and both sexes have pure white abdomens. During winter and spring the whistled *krick'-et* courtship calls of the males can be heard almost as far away as the birds can be seen and often provide the first clue as to their presence. The female has a variety of weak quacking notes and a decrescendo call of about four notes.

Age and Sex Criteria

Sex determination. External characters that indicate a male are vermiculations anywhere, usually on the sides, scapulars, or back. The most distal tertial (adjacent to first iridescent secondary) in males has a black stripe that is sharply delineated, while in females the stripe is blackish to brownish, grading into the basic feather color (Carney, 1964).

Age determination. Notched tail feathers indicate an immature bird, as do tertials that are small, narrow, and rather delicate, with frayed tips. In immatures, middle coverts just anterior to the tertial coverts are often rough and show wear at their edges, and they are usually narrower and more trapezoidal than those of adults (Carney, 1964).

Distribution and Habitat

Breeding distribution and habitat. The North American breeding range of the green-winged teal is similar to that of the American wigeon. On the Aleutian Islands the race *nimia* is a common year-round resident throughout (Murie, 1959; Kenyon, 1961) and is replaced by *carolinensis* on the Alaska Peninsula. The latter form breeds throughout Alaska, except perhaps on the treeless tundra of the Arctic coast, where there are few records of occurrence (Gabrielson and Lincoln, 1959). In Canada the species has an extensive range, from British Columbia and the Yukon on the west to Labrador and Newfoundland on the east,

Fig. 15. Male and female North American green-winged teal.

and northward at least to the tree line. In Newfoundland it is second only to the black duck as a common breeder, and it is also common in Nova Scotia and New Brunswick (Moisan et al., 1967).

In the United States, green-winged teal are common breeders in eastern Washington, are rare in Idaho and Oregon, but are common in extreme northern and northeastern California. Only a few pairs are recorded each year in Utah and Nevada, and they are generally uncommon in the Great Plains states except for the Dakotas, with very limited breeding in Nebraska and Iowa. They are regular breeders in northern

Minnesota, are infrequent in northern Wisconsin, are local breeders in northern Michigan, and are relatively rare in New York. Although regular breeders in Maine and Massachusetts, they are only local to rare breeders in several other eastern states, including Ohio, Pennsylvania, New Jersey, Maryland, and Virginia.

Judging from aerial surveys of the 1970s, the highest continental breeding densities then occurred in the Athabaska Delta, the Slave River parklands, and east of Great Slave Lake. The aspen parklands area of Canada were next highest in density. This would indicate that green-winged teal prefer the wooden ponds of parklands for breeding rather than prairie potholes (Moisan et al., 1967). In 2010 the Traditional Survey of North American waterfowl breeding grounds indicated a population of 3.4 million birds, with 42 percent in the Northwest Territories, British Columbia, and northern Alberta and 27 percent in Alaska and Yukon Territory (Baldassarre, 2014).

Munro (1949) characterized the species' typical nesting habitat as grassland, sedge meadows, or dry hillsides with aspen or brush thickets or open woods adjacent to a slough or pond. Hildén (1964) pointed out that the Eurasian race also prefers to breed on small waters surrounded by woodland, generally does not breed on the eutrophic grassy lakes of open farming country, and avoids open tundra habitats.

Population. Breeding surveys in 2015 indicated a total population of 3.4 million birds, 69 percent above the long-term average of 1.9 million (USFWS, 2016). The estimated total Canadian kill in 2014 was 65,000. The estimated total US kill by hunters averaged about 1.75 million in 2015, or more than half of the total estimated North American population, suggesting that the incompletely surveyed or unsurveyed breeding areas of the species in Alaska may support very large numbers of this species, which is easily overlooked in aerial surveys.

Several hundred birds of the Eurasian form are seen annually during Audubon Christmas Bird Counts at Unalaska Island, Alaska, and a surprising number of Eurasian green-winged teal have regularly been reported among the total annual Canadian hunter kills. Between 1969 and 1993 it was reported in seven years, in numbers ranging from 72 to 5,576 birds, and averaging over 1,000. Considering the difficulties of separating males of these two forms in transitional fall plumage, and the near impossibility of distinguishing females at any time, these numbers might be considered as questionable. The Eurasian race is seen in small numbers more or less annually as a migrant and winter visitor along the Pacific slope in coastal British Columbia (Godfrey, 1986) and at least as far south as Oregon (Gilligan et al., 1994).

Wintering distribution and habitat. The green-winged teal winters along the Aleutian chain (*nimia*) along the coast of southeastern Alaska, south through coastal British Columbia, in the western coastal United States including particularly the Central and Imperial Valleys of California, and southward to central Mexico. In Mexico it is common on both coasts and in the interior but is particularly abundant in Sinaloa and Nayarit (Leopold, 1959). Along the coasts of Texas and Louisiana the species is an abundant winter resident, with an average of 60 percent of the continental wintering population in 1970s Midwinter Surveys occurring in the Mississippi Flyway. Midwinter Surveys from 2000 to 2010 indicated that 44 percent of the birds were in the Mississippi Flyway (90 percent in Louisiana), 27 percent were in the Central Flyway

American green-winged teal, adult male

(96 percent in Texas), 25 percent were in the Pacific Flyway (84 percent in California), and 5 percent were in the Atlantic Flyway (54 percent in North Carolina) (Baldassarre, 2014).

Since most of the Mississippi Flyway birds are produced in western Canada, they evidently migrate south through the Central Flyway and then shift eastward into the coastal marshes of Louisiana (Moisan et al., 1967). It is also thought that, whereas the Central Valley of California receives most of its wintering teal from Alaska, those using the Imperial Valley originate in the Northwest Territories and the Prairie Provinces of Canada.

The preferred wintering habitat consists of coastal marshes, especially those near rice fields in Louisiana and Texas. Open salt water is apparently avoided (Moisan et al., 1967). Stewart (1962) reported that teal prefer creeks and ponds that are bordered by mud flats at low tide. Tidal creeks and marshes of estuarine locations are seemingly preferred over salt marshes. Late fall counts on estuarine bay marshes showed higher usage of fresh or brackish waters, while winter and spring counts indicated a higher use of saltwater marshes.

General Biology

Age at maturity. Green-winged teal probably normally breed at one year of age. Ferguson (1966) stated that 13 of 22 aviculturists reported first-year breeding in American green-wings. The nine reports of two to four years passing prior to breeding are probably a reflection of this species' general reluctance to breed under captive conditions.

Pair-bond pattern. Pair-bonds are reestablished yearly, as with other surface-feeding ducks. There is at least one report of a male in full eclipse remaining with a female and its brood (Munro, 1949).

Nest location. Keith (1961) noted that 22 nests of this species that he found averaged a distance of 65 feet from the nearest water and had an average light penetration at the floor of the nest of only 32 percent, the smallest average figure he reported. He noted that this species and the blue-winged teal had the best-concealed nests of the 12 species he studied. The vast majority (86 percent) of the nests he found were in Baltic rush (*Juncus balticus*) cover, with the rest in mixed prairie and cattails (*Typha*). Girard (1941), reporting on 15 nests, indicated that the average distance to water was 34.2 yards, with a range of 4 to 100 yards. In an Icelandic study of the Eurasian green-winged teal, Bengtson (1970) reported that among 207 nest sites 173 were under shrubs, most of which were less than half a meter high.

Clutch size. Keith (1961) reported an average clutch size of 8.7 eggs for 18 nests. Girard (1941) found that 15 nests had an average clutch of 7.53 eggs. Bauer and Glutz (1968) concluded that 8 to 10 eggs are typical of Eurasian green-winged teal, with normal limits of 5 to 12. Replacement clutches do occur, and average fewer eggs than are in initial clutches (Toft, Trauger, and Murdy, 1984).

Incubation period. The incubation period is probably normally 21 to 23 days, exceptionally to 25 days (Delacour, 1956; Palmer, 1976; Bauer and Glutz, 1968).

Fledging period. Several studies indicate a fledging period of 34 to 35 days in North America (Bellrose 1980), although some longer European estimates have been made (Bauer and Glutz, 1968).

Nest and egg losses. Although his sample size was small, Girard (1941) found that 75 percent of the eggs in 15 nests hatched, and with an average of 5.7 chicks per successful nest. Crows were responsible for some egg losses. Higgins et al. (1992) estimated a 39 percent nesting success rate for 56 North Dakota nests, with known predators including red foxes, striped skunks, badgers, ground squirrels, gulls, and raccoons. Keith (1961) did not calculate a hatching success rate for the 21 nests he found but noted that four nests were deserted, eight were destroyed by skunks and one by an unknown mammal, and at least three hatched. He found that mammalian predation levels were highest in the *Juncus* zone, the preferred nesting cover of green-winged teal.

Fig. 16. Male Eurasian green-winged teal.

Juvenile mortality. Little specific information is available on prefledging mortality, but it is seemingly low. Munro (1949) believed that the high brood survival he observed in green-winged teal was related to the intense brood defense exhibited by females. Moisan et al. (1967) estimated that brood sizes at the time of fledging average from 5 to 7 young. Yocom (1951) found an average brood size of 5.5 young for 27 broods at least two-thirds grown, and Munro (1949) indicated an average brood size of 6.2 young for August broods. However, brood mergers are not uncommon in this species and would influence such counts.

In Iceland, Bengston (1972) estimated a 47 percent rate of duckling mortality prior to fledging. Moisan et al. (1967) estimated a 70 percent first-year mortality rate for teal banded as immature birds.

Adult mortality. An annual adult mortality rate of 50 percent has been estimated for North American green-winged teal (Moisan et al., 1967). This is close to an estimate of 45 percent for Eurasian green-winged teal banded in England (Wainwright, 1967), and an analysis of more than 82,000 banded North American birds that indicated a 55 percent annual survival rate for adult males, 54 percent for adult females, 51 percent for immature males, and 42 percent for immature females (Frank Bellrose, cited in Baldassarre, 2014).

General Ecology

Food and foraging. The small bill of the green-winged teal limits the size of materials it can consume, and plant seeds are apparently an important part of its diet. Martin et al. (1951) listed panic grass (*Panicum*), bulrush (*Scirpus*), and pondweeds (*Potamogeton*) as primary foods, with both seeds and vegetative parts taken in pondweeds. The oogonia of muskgrass (*Chara*) are evidently preferred by green-winged teal but not the "leafy" portions (Munro, 1949). Stewart (1962) noted that the seeds of Olney three-square (*Scirpus olneyi*) and wigeon grass (*Ruppia*), as well as amphipods and gastropods, were the principal foods of 34 birds taken on estuarine bay marshes of Chesapeake Bay, while eight birds from river marshes had consumed seeds of a variety of plants including bulrushes, smartweeds (*Polygonum*), and other aquatics.

In a detailed study of teal food consumption in England, Olney (1963) found that at least during the fall months seeds occurred in nearly all 456 birds examined and represented 76 percent of the total food volume. Most of the seeds ranged from 1 to 2.5 mm in size, with an overall range of 0.5 to 11 mm. Likewise, the mollusks that he found were no longer than 6 mm.

Sociality, densities, territoriality. Green-winged teals are relatively social birds, usually occurring in moderate-sized flocks during both fall and spring. For the most part, however, they do not occur in large flocks. Jahn and Hunt (1964) noted that since teal do not concentrate in refuges but rather remain scattered widely in small flocks, a relatively high hunter kill of this species occurs in Wisconsin.

Estimates of breeding densities are few. In the grassland area of southeastern Alberta, Keith (1961) found a five-year average of 3 pairs (range 2–5) using 183 acres of water on his study area, or a density of 1 pair per 60 acres. Detailed ground surveys in the preferred parkland habitats are not available but no doubt would show higher breeding densities. Atkinson-Willes (1963) commented that the Eurasian race is extremely difficult to study during the breeding season and that it apparently does not occur in high densities anywhere throughout its vast breeding range.

Territoriality is lacking in this species, as indicated by the substantial overlap observed in home ranges of breeding pairs (Frank McKinney cited in Johnson, 1995). Radio-tracking indicated that the home ranges of paired males ranged from 6 to 70 hectares during the egg-laying and incubation periods, and up to 100 hectares when the pair was together or after a nest failure (Johnson, 1995).

Interspecific relationships. Because of its extremely small size and unusually high dependence on small seeds, it is unlikely that the green-winged teal directly competes with any other surface-feeding ducks for food. Rollo and Bolen (1969) noted apparently significant differences in food consumption between green-winged and blue-winged teal during fall in Texas. In comparison with blue-winged teal from the same playa lake, green-winged teal samples showed a higher volume of smartweed (*Polygonum*) seeds and lower amounts of wild millet (*Echinochloa*) and grain sorghum (*Sorghum*). The two species also show considerable differences in wintering areas, migration timing, and preferred nesting habitats. Yocom (1951) observed that

Eurasian green-winged teal, adult males

green-winged teal nest more frequently in the yellow pine (*Pinus ponderosa*) zone of Washington than do the other two nesting species of teals.

Competition for nesting sites is likewise probably negligible, and the green-winged teal is not included in the list of species Weller (1959) listed as parasitizing or being parasitized by other species. Crows (Girard, 1941), skunks (Keith, 1961), and foxes (Higgins et al., 1992) have been noted as important nest predators, although teal nests are usually very well concealed. Bengtson (1972) observed a very low incidence of nest parasitism and listed only minks and ravens as nest predators.

General activity patterns and movements. No specific information on daily activity rhythms appears to be available. Migratory movements have been summarized by Moisan et al. (1967), Johnson (1995), and Baldassarre (2014).

Social and Sexual Behavior

Flocking behavior. During the fall, there is apparently an early southward movement of adult males, while adult females and immatures remain north somewhat longer (Moisan et al., 1967). Jahn and Hunt (1964) found a consistent disproportion of immature males among hunters' kills in Wisconsin, leading them to believe that differential sex migration may occur, with females moving farther south than males. An early spring preponderance of males in sex ratio counts in Washington (Johnsgard and Buss, 1956), as well as in the Netherlands (Lebret, 1950), suggests that females may winter farther south than males. Spring flocks are usually small in size, often consisting of a dozen birds or less.

Pair-forming behavior. Pair formation in the wild probably begins, as it does in the Eurasian race, in early fall and continues through the winter and spring. McKinney (1965) noted that teal he observed in mid-March in Louisiana were virtually all paired. In Austria, about 50 percent of the birds were paired by the end of January, and over 90 percent were paired by the end of March (Bezzel, 1959). However, aquatic social courtship, which begins during September in Austria, does not reach a peak until about the middle of March.

Lorenz (1951–53), McKinney (1965), and Laurie-Ahlberg and McKinney (1979) have all described the pair-forming behavior of this species. In my own limited observations (Johnsgard, 1965) I have not noticed any apparent differences between the male displays of the Eurasian and North American forms (Johnsgard, 1965; see Figs. 12 and 18). The pair-forming displays of the green-winged teal are well known and too numerous (e.g., down-up, head-up-tail-up, grunt-whistle, bridling, etc.) and complex for description here. However, the female's inciting display is very frequent during pair formation, and serves to indicate the female's preference for or pair-bond with a specific male, while the turning-of-the-back-of-the-head ("turn-back-of-head" in McKinney's terminology) display is the typical response of a preferred male to such inciting. Aerial flights are apparently not of special significance in pair formation; McKinney believed that they simply serve to change the location of a courting group, although in many *Anas* species males seem to make a particular effort to fly slightly in front of the female, perhaps thus exhibiting their speculum pattern or their napes.

Copulatory behavior. Mutual head-pumping is the precopulatory display of the green-winged teal. Following copulation the male draws his head backward along the back in a "bridling" display posture (Johnsgard, 1965).

Nesting and brooding behavior. Female teal usually line their nests with a considerable quantity of down and, when leaving, will cover the eggs with the down or other nest lining (Munro, 1949). Females defend their young with remarkable intensity and, if disturbed with a brood on land, will perform distractive movements while dragging one or both wings. When defending a brood on the water they fly or rush about on the water in front of the intruder, often continuing this activity for several minutes while the brood hides in nearby weeds. Munro (1949) documented two females thus jointly defending a merged brood, and he believed that, because of the mother's strong brood defense in this species, there is relatively little mortality of their tiny ducklings.

Postbreeding behavior. Males usually desert their mates about the time incubation begins and may gather in small groups prior to molting. They often move to special molting areas; Hochbaum (1944) noted that although green-winged teal are uncommon breeders in the Delta, Manitoba, area, they poured into the marshes in mid-June and early July. By mid-September migrating teal are common as far south as southern Wisconsin, and populations peak there in mid-October (Jahn and Hunt, 1964). Yet, these small ducks are remarkably cold tolerant and often are seen on Christmas counts as far north as Nebraska, and the Aleutian race is largely resident in those high-latitude islands.

Northern Mallard
Anas platyrhynchos Linnaeus 1758

Other vernacular names. Common mallard, greenhead, green-headed mallard, mallard

Range. Breeds throughout much of the Northern Hemisphere, in North America from Alaska to northern California and east to Ontario and the Great Lakes, with recent breeding extensions into New England. Also breeds in Greenland, Iceland, Europe, and Asia. Winters through much of the breeding range and south to extreme northern Mexico.

North American subspecies. See also the Mexican duck, here called the Mexican mallard, in the Southern Mallards account.

A. p. platyrhynchos L.: Northern Mallard. Range as indicated above, except for Greenland.

A. p. conboschas Brehm: Greenland Mallard. Resident on coastal Greenland, with vagrant birds probably occasionally reaching continental North America (Todd, 1963).

Measurements. *Folded wing: A. p. platyrhynchos:* Males 260–270 mm; females 240–270 mm (Delacour, 1956). *A. p. conboschas:* Males 275–306 mm; females 261–285 mm (Cramp and Simmons, 1977).

Culmen (bill): *A. p. platyrhynchos:* Males 50–56 mm; females 43–52 mm (Delacour, 1956). *A. p. conboschas:* Males 44–51 mm; females 45–52 mm (Cramp and Simmons, 1977).

Weights (mass). Bellrose and Hawkins (1947): Average of 631 adult males, 2.78 lb. (1,261 g); average of 730 immature males, 2.59 lb. (l, 174 g); average of 402 adult females, 2.39 lb. (1,084 g); average of 671 immature females, 2.28 lb. (l,034 g). Nelson and Martin (1953): Maximum male weight 4 lb. (1,812 g); maximum female weight 3.6 lb. (1,631 g). Jahn and Hunt (1964): Maximum male weight 3.81 lb. (1,726 g); maximum female weight 3.81 lb. (1,726 g).

Identification

In the hand. The familiar green-headed and white-collared male in nuptial plumage needs no special attention, but females or immature males may perhaps be confused with other species. Except for the rare Mexican mallard, the presence of a bluish speculum bordered both in front and behind with black and white will serve to distinguish northern mallards from all other North American ducks, with additional criteria being orange-colored legs and feet, a white underwing coloration, and a yellow to orange bill with varying

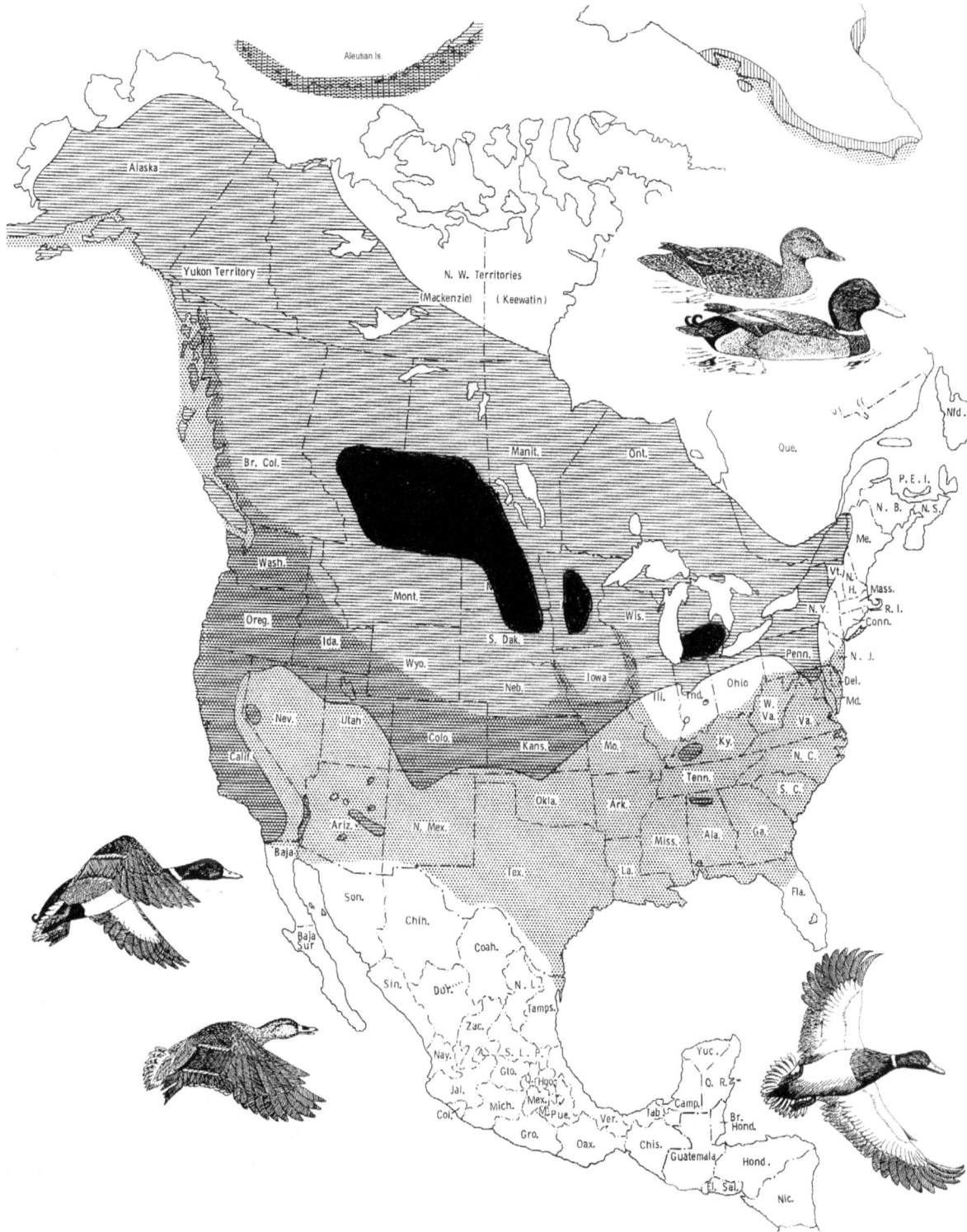

The breeding (horizontal hatching, with denser concentrations inked) and wintering (shaded) range of the northern mallard.

amounts of black present. See the Mexican mallard information (in the Southern Mallards account) for distinction from that species, and the American black duck account for recognition of hybrids. Kirby et al. (2000) also provided criteria for identifying mallards, black ducks, and hybrids using wing plumage traits.

In the field. Mallards are large, surface-feeding ducks that exceed in size all dabbling ducks except the black duck. On the water, the dark green, often apparently black, head color of the male is evident, as are the reddish brown chest and the grayish white sides and mantle, contrasting with the black hindquarters. More than any other dabbling duck, male northern mallards are dark at both ends and light in the middle. Females may be recognized by the combination of their fairly large size and their orange-yellow bill, which is distinctly heavier and more orange than that of a female gadwall. Females also show a definitely striped head, with a dark crown and eye-stripe, contrasting with pale cheeks and a light superciliary stripe. The familiar, loud *quack* of the female is frequently heard, and her call consisting of a series of notes of diminishing volume is also commonly uttered. During aquatic display males utter a sharp whistled note, usually single but sometimes double, that can be heard for several hundred yards. Unlike many other dabblers, this courtship note is not uttered in flight. In flight, the male's immaculate white under wing-coverts contrast with the female's brownish abdomen and upperparts. In males the white of the under wing-coverts is continuous with the whitish sides and abdomen and is terminated in front by chestnut and behind by black. The two white stripes associated with the speculum are also evident in flight.

Age and Sex Criteria

Sex identification. Apart from internal examination or cloacal characters, males older than juveniles usually have some vermiculated feathers present. Wing characters useful for sexing mallards include the vermiculated scapulars, which indicate males. If vermiculations are lacking and the white barring on the greater secondary coverts extends at least to the thirteenth proximal covert, the bird is a female; in males the white does not extend beyond the twelfth secondary covert (Carney and Geis, 1960).

Age determination. *Males:* Juvenal tertials are present until late November. They lack the pearly color of adult tertials and are often frayed and faded. Likewise, juvenal tertial coverts are often frayed, faded, and narrow. Immatures may have light edging on the inner webs of the four most distal primary coverts, and their middle coverts are often frayed, somewhat trapezoidal, and smaller and narrower than those of adults (Carney, 1964).

Females: Frayed or faded tertials or tertial coverts indicate an immature bird, and the two most proximal tertial coverts may lack the white of the anterior speculum bar. Immatures may also have conspicuous light edging on the inner webs of the four most distal coverts, which is lacking or minute in adults, and the middle coverts are narrow and trapezoidal (Carney, 1964). The presence of notched tail feathers indicates an immature bird in either sex.

Fig. 17. Male northern mallard.

Distribution and Habitat

Breeding distribution and habitat. The breeding range of the mallard in North America is extremely broad, probably the broadest of any duck, and comparable to that of the Canada goose. It breeds throughout Alaska, including the Aleutian Islands. In Canada it breeds from British Columbia, Yukon Territory, and the Northwest Territories north approximately to tree line, southeastward across all of western, central, and eastern Canada to southern Quebec and the Gulf of St. Lawrence. Its northeastern breeding limits in eastern Canada extend to James Bay, the southern half of Quebec east to the Gaspé Peninsula, and the Maritime Provinces. Although apparently not yet (2016) nesting on insular Newfoundland, the species' eventual breeding there seems highly likely.

South of Canada, the mallard's breeding range extends broadly across the United States, south to southern California, Arizona, northern New Mexico, eastward across the Great Plains, south to northern Mexico, and east across Texas, the Mississippi River valley, and the southeastern states to Georgia and the Carolinas, and in the Northeast to New England north to Maine. The mallard's invasion of the eastern states and New England as a breeding and wintering species has been a gradual process that may be traced back to the beginning of the twentieth century (Johnsgard, 1961, 1967) and has not yet become stabilized. It finally reached Maine in the early 1950s (Coulter, 1953, 1954) and by the 1970s was a relatively common nesting species there. South of New York and western Pennsylvania the mallard was then distinctly an uncommon breeder, but by the 1970s it had bred in Maryland, South Carolina, Arkansas, Mississippi, and Louisiana. Heusmann (1991) summarized the history of the mallard in the Atlantic Flyway, noting (1988) the significance of urban parks in the rapid expansion of the species into the northeastern United States.

Eastern Surveys covering eastern Canada (extending from Ontario eastward to the Atlantic coast) from 2000 to 2012 averaged more than 396,000 mallards, with 93 percent in Ontario and western Quebec, 4 percent in central Quebec, and the remainder farther east. The 2011 survey included Maine and part of northern New York, and totaled 403,000 mallards. Atlantic Flyway Waterfowl Plot Surveys for the eastern states (excluding Maine) estimated that from 2000 to 2010 there were an average of 731,000 mallards in the flyway (excluding Maine), with 27 percent in New York; 24 percent in Pennsylvania; 8 percent in Virginia; 7 percent each in Maryland, Massachusetts, and New Jersey; 5 percent each in Connecticut and New Hampshire; and the remaining approximate 10 percent scattered throughout the other Atlantic Flyway states (Baldassarre, 2014).

Since the species breeds over such a broad transhemispheric range, it is difficult to separate preferred from acceptable breeding habitats. However, some trends are evident. Hildén (1964) noted that mallards accept waters of almost any kind for breeding, and they will breed in dense woods or on rocky shores as well as around open lakes or on the meadows of grassy lakes. The presence of shallow-water feeding areas and the availability of suitable nest sites appear to be the only critical features. Mallards prefer to nest in fairly dry sites with rather tall vegetation, such as among upland weeds, dry marshes, or in hayfields (Lee et al., 1964a). In forested situations they will sometimes nest in trees or in stumps (Cowardin et al., 1967), but this habitat is not highly preferred by mallards (although it is common for black ducks). Hildén (1964) found mallards breeding on coastal islets covered by grassy or herbaceous growth but not on those that were wooded.

Population. The 2015 North American breeding population was estimated at about 11.6 million birds, 13 percent above the long-term average of 7.7 million (USFWS, 2016). The average annual hunter-kill estimates in the United States during the five years 2004 to 2008 were about 4.62 million birds, and in 2015 was 3.9 million. Estimated total annual Canadian kills of mallards between 1990 and 1998 annually ranged from about 537,000 to 734,000 birds, and in 2015 were estimated at 538,000 birds. In the Atlantic Flyway, the mallard sport-hunting kill fraction has increased from 43 percent of the combined mallard–black duck take in 1964–68 to 80 percent in 2004–08, showing the dual effects of regional mallard expansion and black duck retreat.

Wintering distribution and habitat. Because of their large body size and associated hardiness, mallards are likely to be found wintering anywhere food is available and open water can be found. This includes the Aleutian Islands and the southern coast of Alaska, coastal British Columbia, the coastal states south of Canada, south to extreme northeastern Mexico, and many of the interior states in the southern parts of the United States. As early as the 1970s mallards were wintering regularly as far north on the Atlantic coast as the New England states and extending locally to southwestern Quebec, and rarely to Newfoundland and Nova Scotia. Since then the wintering limits have continued to move northward, both coastally and in the interior, such as to southern Ontario, the southern parts of the Prairie Provinces, and to southern British Columbia.

Northern mallard, adult pair swimming

Stewart (1962) judged that shallow, brackish bays with adjacent extensive agricultural areas represent the optimum habitat for migrant and winter resident mallards in the Chesapeake Bay region. From 50 percent to 86 percent of the fall, winter, and spring population during 1958–59 occurred in this combination of habitats, while estuarine river and bay marshes, coastal salt marshes, and other miscellaneous habitats supported the remainder. Almost no birds were seen on bay marshes having salt water.

General Biology

Age at maturity. Mallards regularly breed at one year of age. This was the opinion of 20 out of 24 aviculturists contacted by Ferguson (1966), and there are many records of wild mallards breeding in their first year of life. First-year females may have somewhat smaller clutch sizes and are less prone to renest than are older females (Coulter and Miller, 1968).

Pair-bond pattern. Mallard pairs are broken and reestablished every year. Once the original mate has left his incubating female, she might re-pair with another mate if renesting is attempted (Sowls, 1955). Lebret (1961) noted several instances of males joining other females after their original mate had begun nesting activities. However, he also mentioned a case in which two birds were known to be paired in each of five consecutive seasons. Reforming of pairs by residential mallard populations may be the rule rather than the exception (Mjelstad, and Sætersdal, 1990).

Nest location. Mallards prefer to place their nests in fairly high vegetation; in one Minnesota study, the average vegetation height at 47 nests was 24 inches, with a range of 10 to 50 inches (Lee et al., 1964b). In a California study (Miller and Collins, 1954), nearly half the nests were located in vegetation 13 to 24 inches tall, with nettle (*Urtica*) and saltbrush (*Atriplex*) apparently being preferred nesting cover. About two-thirds of the mallard nests in this study were concealed on all four sides, and about half were also concealed from above. In a study of mallards in Montana, Girard (1941) found that a third of 267 nests were in tall grasses, and over a fourth were in short grasses.

In a Vermont study, early-nesting mallards often used live conifers or fallen trees for nesting sites, but most later-nesting mallards nested in new or old growth of raspberry or nettle (Coulter and Miller, 1968).

Clutch size. Clutch size data show a surprising amount of variability among different studies, perhaps reflecting the effects of renesting or other influences. Average clutches of about 9.5 eggs have been reported by Lee et al. (1964a), Coulter and Miller (1966), Anderson (1965), and (for early nests) Keith (1961). Clutches averaging 8.5 to 9.0 eggs have been reported by Miller and Collins (1954), Duebbert (1970), Earl (1950), and Hunt and Naylor (1955). Clutches averaging fewer than 8 eggs were reported by Girard (1941) and also by Hickey (1952), who used data from various studies.

Bauer and Glutz (1968) noted similar variations in clutch sizes in European mallards. They established a clear relationship between season and clutch size, with early (March) clutches averaging 10 or more eggs, whereas clutches laid in late May or June averaged from 6.8 to 8.8 eggs. Ogilvie (1964) reported that in England the eggs are laid daily, often with a day's gap during the first 7 eggs.

Incubation period. Incubation under natural conditions averages 28 days, with 2 to 3 days of variation on each side of this mean (Girard, 1941). Ogilvie (1964) reported an average incubation duration of 27.6 days for 51 clutches, with an observed range of 24 to 32 days.

Fledging period. Oring (1968) reported a range of 55 to 59 days in the fledging periods of ten captive mallards, with an average of 56.6 days. This finding is generally in agreement with Hochbaum's (1944) estimate of 49 to 60 days.

Nest and egg losses. A large number of studies have been made on nest success in mallards; Weller (1964) reported that the average of nine studies was 47 percent nesting success, with a range of 13 percent to 85

Northern mallard, adult pair resting

percent. Similarly, Jahn and Hunt (1964), using a variety of studies, estimated that 43 percent of the females succeeded in hatching broods and that the average brood size near fledging was 6.3 young.

Renesting by hens losing their first clutch is common; Coulter and Miller (1968) reported that 53 percent of 32 marked hens were known to renest following nest losses, including females in all stages of incubation at the time of nest loss. In 16 cases the renesting interval varied from 8 to 18 days from the time of nest loss, with no clear relationship between this interval and the stage of incubation at the time of nest loss. The clutch size of 15 renests averaged one egg fewer (9.6 versus 10.6) than the first nests of these females. Keith (1961) estimated that all the unsuccessful females in his study renested, and about a third attempted a second renesting. Humburg, Prince, and Bishop (1978) found that among 11 rematings by females whose nests had been destroyed, 8 of them rejoined their original mates to renest, while 3 found new mates. Similarly, Ohde, Bishop, and Dinsmore (1983) similarly found that 9 of 14 renesting females rejoined their original mates.

Juvenile mortality. Bellrose and Chase (1950) estimated a 55 percent annual mortality rate for juvenile males during their first year after banding. Other estimated mortality rates of as high as 75 percent have also been made (Keith, 1961).

Adult mortality. The annual adult mortality rate for mallards has been estimated at 47 to 48 percent by Hickey (1952) and Gallop (1966), as well as 40 percent (for males) by Bellrose and Chase (1950), and 43 percent (for mallards wintering in England) by Wainwright (1967). Some other estimates of mortality have ranged from 38 percent to 58 percent (Keith, 1961). Since the 1960s a great number of mallard survival rates have been estimated (e.g., Anderson, 1975; Trost, 1987; Smith and Reynolds, 1992). Collectively, they indicate that annual survival rates for adult males range from 62 percent to 68 percent, adult females 54 to 59 percent, juvenile males 48 to 63 percent, and juvenile females 46 to 61 percent (Baldassarre, 2014).

General Ecology

Food and foraging. One of the mallard's significant foraging characteristics is its ability to utilize agricultural grain crops as well as natural aquatic foods, depending on their relative availability. Important natural foods include wild rice (*Zizania*), pondweeds (*Potamogeton*), smartweeds (*Polygonum*), bulrushes (*Scirpus*), and a large variety of other emergent submerged plants (Martin et al., 1951). The proportion of animal material in their diet is usually under 10 percent and is probably highest during summer. Farm crops that are often heavily utilized include corn, sorghum, barley, wheat, oats, and almost any other grains that might be available.

Girard (1941) noted that in Montana field-feeding by mallards began in mid-August. The birds begin to congregate in groups about 2:30 p.m., and all leave their water areas between 3:00 p.m. and 6:00 p.m. They often feed all night and return to water between 7:30 a.m. and 8:30 a.m. During the hunting season, the feeding schedule is somewhat modified, and the birds both leave to feed late in the afternoon and return earlier in the morning, thus avoiding exposure to hunters. During winter in Montana, the birds usually remain on the water all night. Their chief food in Montana is wheat, although they also consume barley, oats, and rye.

Winner (1959) made a similar study of field-feeding in mallards and black ducks. He found that afternoon feeding flights of mixed mallard and black duck flocks began from 9 to 205 minutes before sunset, with flight being initiated earlier as the flock size and/or percentage of mallards in the population increased. Winner found no clear relationship between flight initiation and temperature, absolute light intensity, or the time at which legal shooting terminated. Bossenmaier and Marshall (1958) noted that mallards and pintails left on their morning feeding flights at daybreak, or about 30 minutes before the geese left on their flights, and sometimes would be back on the lake before the geese had left. Bossenmaier and Marshall observed no overnight foraging and noted that feeding flights occurred in all types of unfavorable weather, including fog and blizzards.

Stewart's (1962) study of the foods of 85 mallards from the Chesapeake Bay region indicated the foods there varied locally among birds collected in estuarine bays, estuarine river marshes, estuarine bay marshes, and river bottomlands. In the estuarine areas, seeds of shoreline, emergent, or submerged plants (*Scirpus, Polygonum, Sparganium, Potamogeton*, etc.) were prevalent, as were the leafy portions and rootstalks of submerged species. In Louisiana, mallards have made increasing use of rice or other plants associated with the culture of rice in recent years, and in one study more than 90 percent of the wintering mallards were located in or near the rice-growing area (Dillon, 1959).

Like most other surface-feeding ducks, mallards will sometimes dive to obtain their food (Kurtz, 1940; Kear and Johnsgard, 1968), although tipping-up is the usual manner of foraging. When foraging in grain fields, mallards can consume surprising amounts of grain, which in one study averaged about seven ounces per bird per day, assuming two feedings each day (Bossenmaier and Marshall, 1958).

Sociality, densities, territoriality. Shortly after mallards have completed their prenuptial molt into their winter plumage, social courtship starts and flocks of both sexes begin to form. Large flocks are facilitated where feeding in grain fields occurs, since mallards tend to move back and forth between their resting and foraging areas in fairly large flocks. Winner (1959) noted that mixed winter populations of mallards and black ducks on a 940-acre reservoir ranged in size up to about 8,000 birds, with up to several thousand feeding in a single cornfield.

Flock sizes remain fairly large throughout winter and gradually tend to break up as paired birds separate from flocks containing unmated males likely to harass females. Breeding densities vary greatly in different habitats but are generally not extremely high. Drewien and Springer (1969) noted an average density over a 16-year period of 6.7 pairs per square mile in prairie pothole habitat in South Dakota. Stoudt (1969) reported a 15-year average density of 28 pairs and nine broods per square mile in a Saskatchewan study area, and noted that four other study areas have had peak mallard densities of from 9 to 54 pairs per square mile.

Duebbert (1969) reported a nest density of 24 nests on a 125-acre field, although only 17 pairs were observed on the four-square-mile study area. He suggested that some female mallards may have flown 3 to 5 miles to this area of prime nesting cover. Drewien and Fredrickson (1970) estimated that 78 mallard nests were on a 19-acre South Dakota island in 1967, and 60 nests in 1968. In 1967 Drewien and Fredrickson found an average distance between nests of 34 feet, with a range of 7 to 150 feet. Favored nesting cover in the form of tall nettles and protection from predators evidently had been responsible for this unusual density.

The existence of true territoriality in mallards as well as in most other surface-feeding ducks is doubtful. Dzubin (1955) concluded that mallards do not defend a rigid area and that apparent territories may overlap with those of other pairs of mallards. Additionally, the female is defended outside the limit of the "territory." This and other studies make it clear that the female, rather than a specific area, is the male's focus of defense, and a territory in the classic sense of a defended area does not exist (Raitasuo, 1964). Hori (1963) suggested that aerial chases in mallards are more a reflection of a tendency toward polygamy (if not promiscuity) than evidence for territoriality, and McKinney (1965) believed that such chases served as a mechanism for dispersion of pairs. Thus the term "home range" is more properly applied to an area within which a breeding pair of ducks remains but which is not defended as such.

Interspecific relationships. The close evolutionary relationships existing between the mallard and the American black duck (Johnsgard, 1959, 1961c) suggest that interspecific competition between them may be significant in their considerable area of present overlap. Mixed courtship groups of these two species indicate that some interspecific competition for mates does exist, although the rate of mixed pairing and subsequent hybridization is well below that of nonselective mating (Johnsgard, 1967a).

Coulter and Miller (1968) found that nest sites selected by mallards and black ducks were quite similar, although they did not analyze the relative attraction of these two species to different habitat types. On islands in Lake Champlain, mallards showed a higher rate of use of dead herbaceous plants, such as nettle, and tree boles, crotches, and stubs for nest sites, whereas black ducks had a higher usage rate of fallen limbs or logs and dead treetops. Coulter and Miller believed that such use of wooded islands by black ducks was common only where sedge-meadow bogs, their preferred habitat, were not nearby. In contrast, the mallard prefers nesting on typical grassland marsh habitats and likewise is not attracted to wooded habitats (Johnsgard, 1959). However, both species can and will use stumps and trees for nesting in special situations (Cowardin et al., 1967).

Besides competition with other ducks, mallards have the usual number of egg and duckling predators with which to contend. These include skunks, raccoons, coyotes, hawks, crows, and many other birds and mammals, as well as snakes, snapping turtles, various other predatory reptiles, and some fish.

General activity patterns and movements. Like other surface-feeding ducks, the mallard is largely diurnal and has a polyphasic pattern of activities that recur throughout the day that is in part related to temperature, wind, light, and other environmental variables (Raitasuo, 1964). Some overall patterns can, however, be detected in the birds' behavior patterns. Girard (1941) noted that during April observations most resting occurred during midmorning and midafternoon hours, mating and fighting activities were mostly seen in the morning, foraging in water was seen both during morning and afternoon, and foraging adjacent to the shore or on land near shore was mostly seen in late afternoon. Field-foraging flights typically occur close to sunrise and sunset.

Winner (1960) studied movements of marked mallards and black ducks during late fall and winter on O'Shaughnessy Reservoir in central Ohio. Of 62 individually marked mallards, Winner found that their stopover period on the reservoir lasted up to 18 days, with an average of 3.4 days. Ducks left the reservoir under all weather conditions, but the two largest decreases he observed occurred during weather conditions characterized by an overcast sky, falling barometric pressure, relatively constant temperature, and southerly winds.

Social and Sexual Behavior

Flocking behavior. The mallard's adaptability to field-feeding in grain fields and its large size and associated hardiness are in large measure responsible for its ability to winter relatively far north in the grain-growing belt of North America, spending the night on large lakes or reservoirs and feeding in adjacent grain fields. Jahn and Hunt (1964) noted that during October mallards would readily fly 15 to 25 miles from an aquatic

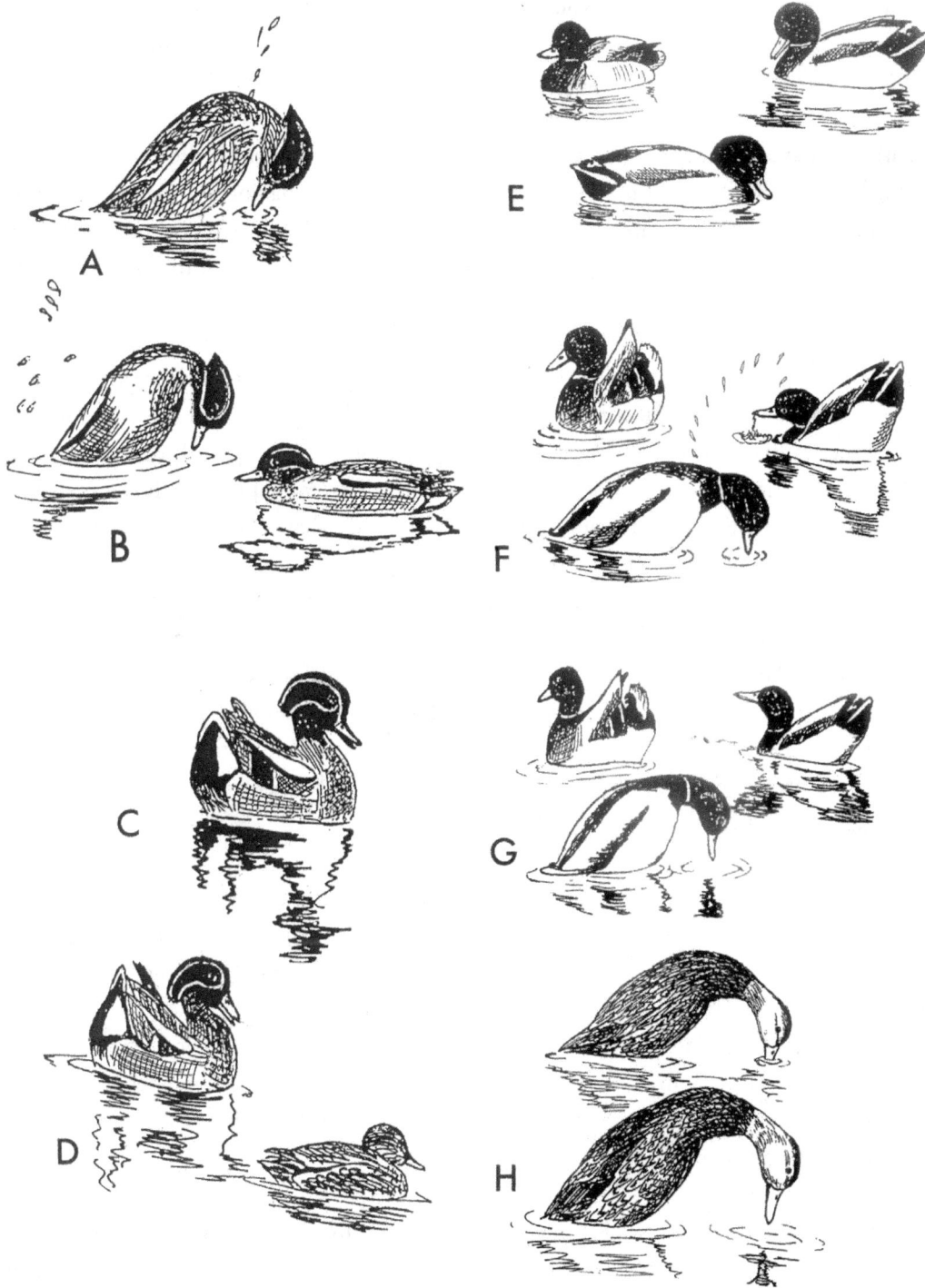

Fig. 18. Sexual behavior of green-winged teal (A–D), northern mallard (E–G), and Florida duck (H), including (A–B) grunt-whistle, (C–D) head-up-tail-up, (E–G, left to right) head-up-tail-up, grunt-whistle, and down-up, and (H) grunt-whistle.

concentration site to feed on corn and would remain in agricultural areas of Wisconsin on into winter. Even as far north as North Dakota, mallards in substantial numbers now (2016) winter on the Garrison Dam reservoir (Lake Sakakawea) and other large reservoirs.

Pair-forming behavior. In September, as juvenile mallards begin to assume their first winter plumage and as adult birds are regaining their nuptial plumages, pair-forming behavior is initiated. It is apparent that if members of previous pairings locate one another, they will reestablish their pair-bonds without any special ceremonies, and this accounts for the moderate number of paired birds seen in early fall before social display begins in earnest (Lebret, 1961).

Shortly after about 90 percent of the males have assumed their nuptial plumages, social display reaches a peak of activity and continues at a relatively high level through the winter and spring (Bezzel, 1959). Before the end of the year, at least 90 percent of the females are already paired in many regions; thus, it is apparent that a substantial amount of "courtship" display must go on among birds that are already apparently paired. This display may help serve to strengthen pair-bonds, but more probably it channels aggressive tendencies toward other males into a ritualized pattern of behavior that reduces actual fighting and facilitates the maintenance of the flock (Lebret, 1961).

Although the complex aquatic courtship displays (see Fig. 18) of males must influence—in ways still uncertain—mate choice among females, the actual pattern of pair formation between individual birds is much less conspicuous. In large part it evidently consists of females inciting "chosen" males against others and of the associated responses of such males, which may include hostile responses toward the indicated "enemy" as well as a ritualized turning-of-the-back-of the-head display toward the female. Mutual drinking behavior and ritualized preening displays by the male toward the female are other important aspects of pair formation in mallards (Johnsgard, 1959, 1965).

Copulatory behavior. Copulation in mallards is preceded by mutual head-pumping, which may be initiated by either sex. As treading is completed, the male releases his grip on the female's neck, draws his head backward along the back in a "bridling" movement as a whistle is uttered, and then swims rapidly around the female in a "nod-swimming" display (Johnsgard, 1965).

During late spring, especially as females are beginning to nest and are no longer so closely guarded by their mates, a great deal of raping behavior is characteristic of mallards. These rapes are mostly performed by unmated males, but males that have recently deserted their incubating mates may also participate in such behavior to some extent.

Nesting and brooding behavior. McKinney's (1953) study of incubation behavior in mallards is unusually complete, and other biologists have made less intensive observations. Once the female begins incubation, she normally leaves the nest only twice a day to feed. Girard (1941) noted that about two hours are taken each day for foraging, usually between 6:30 a.m. and 8:30 a.m. and again in the late afternoon. Coulter and Miller (1968) noted, however, that considerable variation in the feeding period occurred, and McKinney (1953) reported that feeding periods usually lasted only 30 to 60 minutes.

When on the nest, the female may change her position at a rate averaging once every 35 minutes (McKinney, 1953). The bird then typically rises, preens or tugs at her breast feathers, turns a varying amount, and settles back down on the eggs. Then she "paddles" with her feet in a manner that helps to turn the eggs. Finally, she pats in the nest edge with the underside of her bill and pulls nesting material in toward the nest. Down gradually accumulates in the nest by the preening and tugging action of the female and may be quite abundant by the time of hatching.

Mallard eggs require about 30 hours to complete pipping (Girard, 1941), and most of the eggs hatch during daylight hours (Bjarvall, 1968). The first night after hatching is typically spent in the nest, and the family leaves the nest the next morning, usually before 10:00 a.m. (Bjarvall, 1968). The female normally looks after her brood for most of the eight-week period required for the young to attain flight. However, several instances are known in which a female has laid a second clutch after successfully hatching an earlier one, and in at least two of these cases part of the original brood was still alive at the time the female had started her second clutch (Bjarval, 1969).

Postbreeding behavior. Male mallards desert their incubating females at varying times, from as early as the start of incubation until as late as the third or fourth week of incubation. However, there is a still undetermined period following desertion of the female during which sexual vigor is retained (Johnson, 1961), and such males may for a time be of significance in facilitating renesting or in mating with other females.

Males about to lose their flight feathers often gather in flocks of several hundred to several thousand birds, loafing on beaches and feeding in marshes, sloughs, meadows, and the like. However, with the loss of flight ability, the males become extremely secretive and are rarely seen (Hochbaum, 1944). Following a flightless period of about 24 to 26 days (Boyd, 1961), the males again begin to gather in conspicuous places. Females usually do not begin their wing molt until they have abandoned their well-grown broods, and thus the peak of their flightless period occurs more than a month after that of males.

Southern Mallards:
Mexican, Florida, and Mottled Ducks
Anas (p.) diazi, Anas (p.) fulvigula, and *Anas (p.) maculosa*

Note: The parenthetic *(p.)* indicates that this taxon is here considered to be part of the common mallard super-species *Anas platyrhynchos.*

Other vernacular names. *A. (p.) diazi:* dusky mallard, New Mexican duck; *A. (p.) maculosa:* summer black duck, summer mallard, western Gulf Coast mottled duck.

Range. Southern mallards currently exist as three separate, largely residential, populations. One (*fulvigula*) is resident in peninsular Florida, with recent range extensions to coastal South Carolina and Georgia. A second population (*maculosa*) breeds along the Gulf coast from the Mississippi Delta to central Veracruz, wintering over most of the breeding range but probably undergoing some seasonal movements. The third population (*diazi*) is now largely influenced by mallard introgressive hybridization, and its northern remnants are limited to a breeding range in the Rio Grande valley of southern New Mexico, extreme southwestern New Mexico, and adjacent Arizona. It also occurs locally in its historic range wetlands from Chihuahua, Durango, northern Jalisco, and the central highlands of Mexico south to the Trans-Mexican volcanic belt, where the mallard influence so far is slight. Wintering in all three populations occurs through much of the breeding range, but there is probably a limited movement out of the northernmost breeding areas in the United States, and an expanded distribution from the central Mexican highlands into eastern Mexico (Howell and Webb, 1995).

Species and subspecies. *Anas platyrhynchos diazi* Ridgway: Mexican Mallard. Resident in New Mexico, Arizona, and Mexico, as indicated above; the current population is largely limited to Mexico, from Chihuahua south to central Mexico.

    A. (p.) fulvigula Ridgway: Florida Duck. Resident in Florida, South Carolina, and Georgia.

    A. (p.) maculosa Sennett: Mottled Duck. Resident on the Gulf coast from Mobile Bay west and south through coastal Texas and eastern Mexico to Veracruz. This taxon was not recognized by Delacour (1956), but Johnsgard (1959, 1961c) concluded that it is probably a valid subspecies, and recent molecular data also support racial recognition (McCracken et al, 2001).

Measurements. *Folded wing* (Kear, 2005): *A. p. diazi:* Male (52) 260–282 mm, average 269.9 mm; female (13) 232–268 mm, average 253.4 mm.

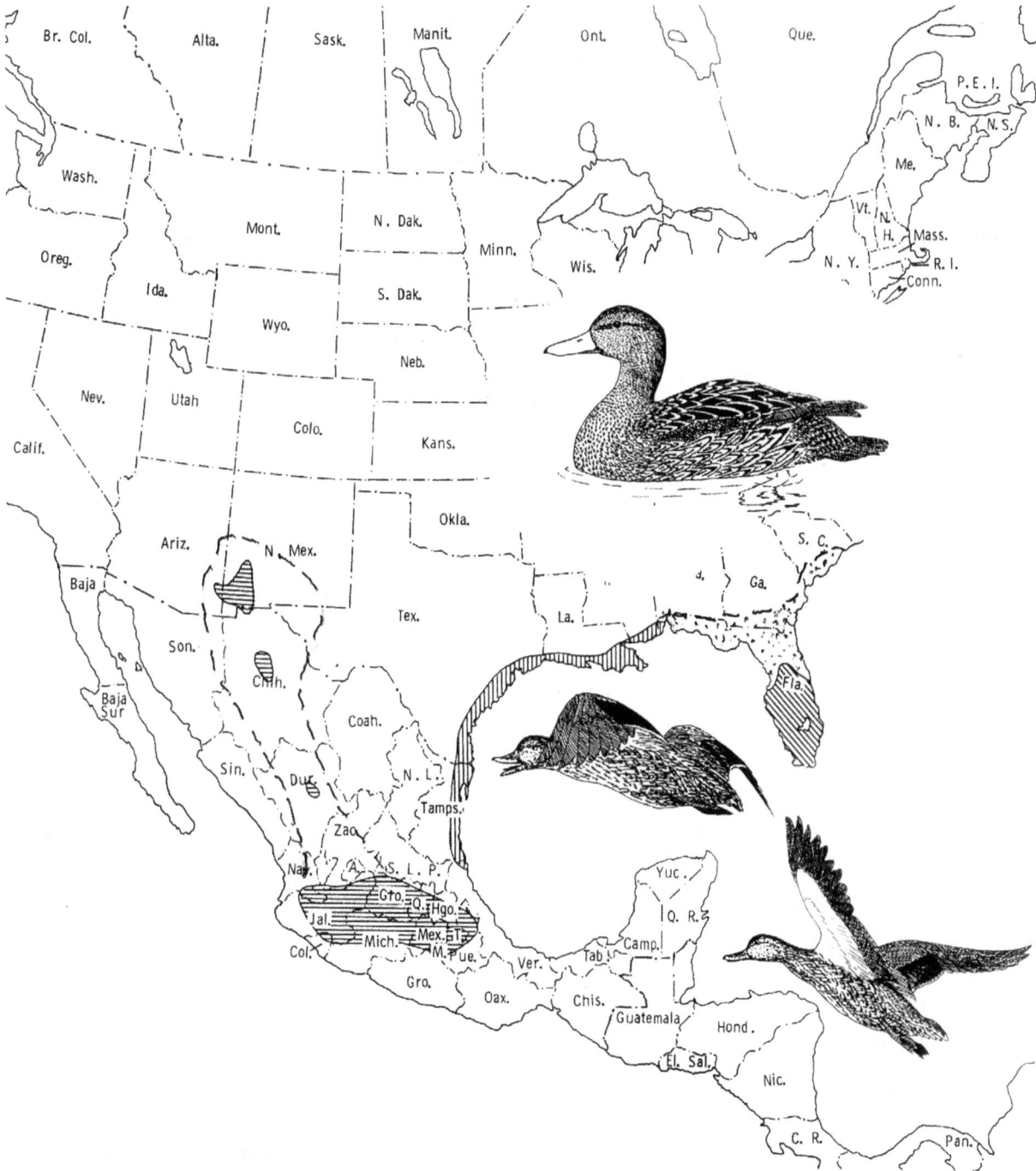

The breeding and wintering ranges of the Mexican mallard (horizontal hatching), Florida duck (diagonal hatching, with acquired range stippled), and mottled duck (vertical hatching, with acquired collective range of the Florida and mottled ducks stippled).

A. (p.) fulvigula: Male (21) 237–264 mm, average 252.0 mm; female (13) 248–264 mm, average 256.0 mm.
A. (p.) maculosa: Male (13) 248–264 mm, average 252.0 mm; female (9) 222–245 mm, average 234.0 mm.
Culmen (bill) (Kear, 2005): *A. p. diazi:* Male (13) 51.1–55.6 mm, average 53.3 mm; female (13) 45.4–52.7 mm, average 50.3 mm.
A. (p.) fulvigula: Male (21) 51.1–57.8 mm, average 54.6 mm; female (13) 49.1–52.1 mm, average 50.4 mm.
A. (p.) maculosa: Male (13) 50.1–63.1 mm, average 54.9 mm; female (9) 47.8–54.95 mm, average 51.7 mm.

Weights (mass). *A. (p.) diazi:* Males 2.13–2.36 lb. (960–1,060 g); females 1.8–2.17 lb. (815–990 g) (Leopold, 1959). Males 647–1,243 g, average of 52, 1028 g; females 647–1,257 g, average of 48, 908 g (Scott and Reynolds, 1984).

A. (p.) fulvigula: Males average of 30, 2.19 lb. (994 g), maximum 2.81 lb. (1,273 g); females average of 11, 2.19 lb. (994 g), maximum 2.5 lb. (1,132 g) (Beckwith and Hosford, 1955).

A. (p.) maculosa: Males average of 26, 2.27 lb. (1,028 g.); females average of 10, 2.04 lb. (927 g) (Hoffpauir, 1964).

Identification

In the hand. *Adult males* of all three taxa are generally similar to females of the common mallard, especially *diazi*, which, however, is more heavily streaked and spotted with brown on the underparts and has an unspotted yellow bill with (usually) black nails.

Males of *maculosa* and *fulvigula* are generally darker and tawnier, with yellow to yellowish orange bills, black nails, and a black mark near the base of the upper mandible. They also lack a definite white bar on the greater secondary coverts since this area is suffused with tawny.

Females of all three populations are virtually identical to the males except for bill coloration, but they tend to have wider buffy edging on their body feathers and thus have a slightly lighter overall body plumage.

Females of *diazi* can be distinguished from female common mallards by one or more of the following traits: (1) the upper tail-coverts are darker, with no patterning along the quill and with narrower light margins; (2) the outer tail feathers are darker, with little or no white present; (3) the under tail-coverts are dark brown with a lighter edging instead of being white with a central brownish stripe; (4) the small under wing-coverts are barred with brown; (5) the bill is darker, shading anteriorly to olive green with very little orange near the base; (6) the tertials are overlaid with a greenish cast; (7) the speculum is more greenish and has a reduced white border; and (8) the breast feathers usually are a darker brown, varying in pattern from three separate spots to a merged fleur-de-lis (Huey, 1961).

Females of *fulvigula* and *maculosa* tend to be even darker than those of *diazi* and may have a more purplish speculum without a definite white anterior border.

In the field. Birds of all three populations look very much like female common mallards in the field but average variably darker in their plumage tones. The major difference is that both sexes have a yellow or olive bill color with little or no dark spotting present, and when in flight the birds exhibit little or no white on their outer tail feathers. The body tones of *diazi* are sometimes only slightly darker than those of female common mallards, but females of *fulvigula* and *maculosa* are distinctly more tawny. These latter types also lack a definite white bar in front of the speculum. Female hybrids between common mallards and black ducks are very similar to parental females and are probably impossible to distinguish in the field. Such hybrids retain a small but distinctive white or grayish white bar on the greater secondary coverts, which would help to separate them from either Florida or mottled ducks, the only forms likely to be encountered where hybridization between common mallards and black ducks is most prevalent.

Age and Sex Criteria

Sex determination. Adult males have a bill that is entirely yellow, except for a black nail (sometimes yellow in *diazi*) and a black spot near the back of the upper mandible (lacking in *diazi*). Females have a more olive-colored bill (sometimes orange basally in *diazi*) that grades to olive green toward the top, or has limited black spotting on the sides and top (usually absent in *diazi*). Internal examination may be required in the case of immature birds.

Age determination. Probably the criteria for age determination mentioned in the account of the black duck may be applied to these populations as well.

Taxonomic Comments

The subspecies *diazi* was regarded as a separate species by the American Ornithologists' Union from 1957 (5th edition of the *AOU Check-list*) until 1983 (6th edition of the *AOU Check-list*), when it was merged with *platyrhynchos*, and *platyrhynchos*, *fulvigula*, and *rubripes* were acknowledged as appearing to constitute a superspecies. The northern taxon (*novimexicana*) was not recognized by Delacour (1956), Johnsgard (1961c), or Aldrich and Baer (1970).

In Texas wild hybrids between the mottled duck and northern mallard have been extensively reported (Nelson, 1980; Benson and Arnold, 2001; Lockwood and Freeman, 2014). As an interesting historical background event, during the early part of the twentieth century J. C. Phillips (1921) experimentally raised some captive hybrids between the northern mallard and a "Florida duck," which was actually a mottled duck from Louisiana. The F1 generation hybrids were fully fertile, and the F2 generation offspring ranged from the extreme mallard phenotype to the mottled type (Johnsgard, 1961c), suggesting that even a century ago it was recognized that a very close genetic relationship exists between the two (Johnsgard, 1961c).

Nearly a century later and using molecular data, Kerr et al. (2007) reported that the DNA "barcodes" they used for recognizing different species were unable to distinguish the mottled duck, American black duck and northern mallard from one another. McCracken, Johnson, and Sheldon (2001) recently also concluded that the mottled duck is a very close relative of the American black duck, sharing a fairly recent common ancestry with this species, and there is some other evidence supporting this view (Johnson and Sorenson, 1999; McCracken et al., 2001; Kulikova et al., 2004). Lavretsky (2008), using mDNA and 17 nuclear loci, found that Florida and Gulf coast populations were differentiated from one another and other taxa studied, but northern mallards, American black ducks, and Mexican ducks were not significantly differentiated.

The absence of a viable species-level definition of *fulvigula*, based on (1) the genetic similarities of it with both *platyrhynchos* and *rubripes*, and (2) the criterion of reproductive isolation from *platyrhynchos* (see below for more details) effectively separating *fulvigula* from the northern mallard, makes a biologically defensible species-level distinction impossible. In this book I have compromised with traditional nomenclature and classified both *maculosa* and *fulvigula* as members of the mallard superspecies by designating them as *Anas (p.) fulvigula* and *A. (p.) maculosa*. Following AOU precedence, I designated only *diazi* as a subspecies of the common mallard (*A. p. diazi*). Yet, as noted by the AOU (1998, p. xiv), "essential (lack of free interbreeding) rather than complete reproductive isolation continues to be the fundamental operating criterion for species status." By that criterion it is obvious that both *fulvigula* and *maculosa* should both also be downgraded to subspecies status, as I did in 1961, 1967, this book's 1975 edition, and later more formally (Johnsgard, 1979, p. 469). It is now painfully apparent that *rubripes* probably also will eventually require the same treatment should current hybridization rates continue to increase.

Distributions and Habitats

Mottled and Florida ducks. The "Florida duck" (*fulvigula*) was historically limited to the peninsular portion of Florida, with the population's northern limits at about Cedar Key, Gainesville, and Daytona. An estimate of 50,000 birds was made during the fall of 1966 (Stieglitz and Wilson, 1968), and in 1984 Johnson et al. estimated a similar 1977–80 average fall population of 67,000. More recent surveys have suggested a spring Florida population of about 53,000 birds, but these have included at least some northern mallards and mallard–Florida duck hybrids (Bielefeld 2008; Bielefeld et al., 2010).

Northern mallards have been increasing recently in Florida at least partly as a result of releases by hunting organizations. The Florida duck's population has been dramatically impacted by the resulting hybridization (Bielefeld et al., 2010), which is likely to cause increasing problems in maintaining genetic purity for the Florida duck. Hybridization rates between Florida ducks and northern mallards recently varied geographically from 0 percent to 24 percent, and up to 10.9 percent of the Florida ducks have been identified as hybrids (Williams et al., 2005). Additionally, between 1975 and 1982 mottled ducks from Texas and Louisiana were introduced into coastal South Carolina and later spread south to coastal Georgia. Hybridization with northern mallards has already been detected in South Carolina (Weng, 2006). This expanded *maculosa* population may have also already been in contact with the Florida duck's population in Florida, since some of the Georgia birds have since spread into northern Florida and begun to hybridize (Weng, 2006).

Mottled ducks (*maculosa*) from Louisiana (which are now classified as part of the Western Gulf Coast

Fig. 19. Male Florida duck, two-wing stretch.

Population of mottled ducks, distinguishing it from the Florida Population) currently (2016) occur at least as far east along the Gulf coast as Mobile, Alabama. In Louisiana the birds are fairly common in coastal areas, especially in Louisiana's rice-growing regions. As early as the late 1960s the total average winter population was estimated at 40,000 to 70,000 birds.

These ducks still occur fairly evenly over the marshes of southeast and southwest Louisiana, especially in the salt and brackish marshes along the coast, and have benefitted from the marsh habitat changes brought about by recent hurricanes. The Louisiana population is continuous with the Texas population, which extends from the Louisiana-Texas border to the Mexico border. Summer populations over a 26,000-square-mile area in Texas were estimated at 20,000 birds in 1952, which then occurred from the coast inland from 50 to possibly 100 miles (Singleton, 1953). More recent (1994–95) spring counts along the Texas coast have produced estimates of more than 100,000 breeding pairs (Ballard et al., 2001).

Besides the historic breeding range of *maculosa*, there are also extralimital records of mottled ducks breeding at Cheyenne Bottoms Waterfowl Refuge in Kansas between 1963 and 1977, and occurrence records from eight other Kansas counties (Thompson et al., 2011). There is also at least one record of *maculosa* from Oklahoma and records of banded birds recovered from as far away as Wisconsin and New Jersey (Moorman and Gray, 1992).

Breeding habitat preferences for the mostly coastal-dwelling populations of mottled ducks have not been carefully analyzed. Engeling (1949) described the preferred habitat of Texas mottled ducks as salt marshes, coastal prairies, bluestem meadows, and fallow rice fields. Nesting is usually in open prairies, and later birds move to rice fields and marshes. Beckwith and Hosford (1957) found Florida ducks nesting near Lake Okeechobee, Florida, on a relatively flat habitat having about 65 percent of the area in wet prairies, seasonal marshes, and sloughs; 13 percent in ponds, most of which were shallow; 1.3 percent in sawgrass (*Mariscus*) marsh; and the remaining approximately 30 percent in varius terrestrial vegetation.

Mexican and New Mexican ducks. In addition to the Florida and Gulf coast mallard-like populations described above, there is also an interior population of mallard-like ducks (*diazi*) that had a historic breeding range extending from southeastern Arizona and central northern New Mexico south to central Mexico (Johnsgard, 1959; Aldrich and Baer, 1970; Webster, 2006). The subpopulation breeding north of the Mexican border and previously known at the New Mexican duck ("*novimexicana*") has been greatly reduced in population and its genetic identity has been heavily diluted by extensive hybridization with northern mallards throughout New Mexico (Hubbard, 1977), but Webster (2006) stated that the small population still breeding in Arizona is not obviously influenced by northern mallard traits. Furthermore, breeding by *diazi*-phenotype birds still occurs rarely to uncommonly in southern Texas. There it is restricted to the Rio Grande watershed from Webb County to Hidalgo County, but there are occurrence records from Crosby, Lubbock, Midland, and Swisher Counties (Lockwood and Freeman, 2014). There are also some older specimen records from Nebraska and Colorado.

One early habitat study of *diazi* in New Mexico was by Lindsey (1946). He located four nests, all in meadows or lowlands containing three-square (*Scirpus americanus*), salt grass (*Distichlis*), rush (*Juncus balticus*), sedge (*Carex*), or barley (*Hordeum*). Leopold (1959) noted that nearly all the habitats where he observed *diazi* contained some cattail (*Typha*) or tule (*Scirpus*) marsh, and that this type of wetland seemed to represent their preferred habitat.

Hubbard (1977) noted that during the breeding season *diazi*-like phenotypic birds in New Mexico strongly preferred native riparian and pond habitats, but Scott and Reynolds (1984) noted that in Mexico the birds apparently have become adapted to the many large irrigation and grain agricultural systems that have developed throughout the Mexican highlands. These habitats should persist into the future, helping to ensure the continued existence of *diazi*.

The Mexican component of *diazi* extends from the Texas-Mexico border south to central Mexico and probably represents about 98 percent of the taxon's overall population. There is substantial phenotype variation in the overall Mexican population, with the northern populations influenced by northern mallard phenotypes (Scott and Reynolds, 1984). In spite of these variations, Scott and Reynolds anticipated no danger

Mexican mallard, adult male

of overall genetic swamping by northern mallards, nor any special concern for the future of *diazi*, partly because of the reputed "natural wariness" of these birds, and also because of the relatively small numbers of northern mallards currently reaching central Mexico.

Populations. The combined 1990s populations of *fulvigula* and *maculosa* probably consisted of about 56,000 birds in Florida and 500,000 to 800,000 in Texas and Louisiana, with substantial year-to-year variations that might have been the result of drought and/or overhunting (Moorman & Gray, 1994; McCracken et al., 2001). The average annual hunter-kill estimate in the United States of "mottled ducks" (mottled ducks plus Florida ducks) during the years 2004 to 2010 varied from about 70,000 to 80,000 birds. It ranged from 67,000 in 2011 but declined progressively to 41,000 in 2015, averaging over that period at about 52,400.

Scott and Reynolds (1984) estimated the spring 1978 population of *diazi* in Mexico at 55,000 birds. Pérez-Arteaga, Gaston, and Kershaw (2002) surveyed Mexican mallard populations periodically from 1960 to 2000, during which time the largest single count was nearly 50,000 in 1988. A stable or slightly positive long-term population trend has been occurring, although the comprehensive federal surveys over the long-term have reported 10,000 to 20,000 birds. There is no information on the extent of hunter-kill in Mexico.

General Biology

Age at maturity. Six of seven aviculturists responding to a questionnaire by Ferguson (1966) said that Florida ducks mature their first year. Beckwith and Hosford (1967) also noted that reproductive maturity occurred in Florida ducks during their first year of life.

Pair-bond pattern. Observations on social display are relatively few but indicate that the period of pair formation, sexual behavior, and the type of pair-bond formed differ in no substantial way from that of mallards or black ducks (Johnsgard, 1959, 1961c). Singleton (1953) noted that the maximum number of paired birds seen was during March and the minimum was during August, when only 4 percent appeared to be paired.

Stieglitz and Wilson (1968) raised the possibility that, in the Florida population at least, the pair-bond may be virtually permanent, since mated pairs were seen all year and males seemed to be absent only during the brood-rearing period. Engeling (1951) mentioned that two birds banded as a pair in January of 1949 were shot together in January of 1950, indicating the maintenance of a pair-bond through one brooding season. Other indications of possible long-term bonding have been made in New Mexico (S. Williams, in Baldassarre, 2014), and Mexico (Williams, 1980).

Nest location. Stutzenbaker (1988) found most of 315 Texas nests in dense stands of cordgrass (*Spartina*), and nearly all of them were hidden from above by overhanging cordgrass as well as being somewhat elevated above the ground by a bed of cordgrass support. More than half of 39 nests found by Finger et al. (2003) also were located in cordgrass, and most had an overhead canopy of vegetation.

In a Florida duck study (Stieglitz and Wilson, 1968) it was found that paspalum (*Paspalum*) was the dominant plant at 55 percent of 88 nests, and broom sedge (*Andropogon*) dominated at 18 percent. Cover height at nest sites averaged 34 inches and ranged from 6 to 96 inches. The nests averaged a distance of 28 feet from water, and almost 80 percent were 10 to 40 feet from water.

Lindsey (1946) described several *diazi* nests in New Mexico. One was in a low *Scirpus-Distichlis* meadow, one in a moist *Distichlis* meadow, one in a *Juncus* meadow, and one in a growth of *Carex* and scattered *Hordeum*. Their placement ranged from almost immediately beside water to a distance of 0.1 mile from the nearest water. Beckwith and Hosford (1957) noted that most of the five nests they found in Florida were located near water and that three were in tomato fields.

Clutch size. Singleton (1953) reported that 108 nests of *maculosa* in Texas averaged 10.4 eggs per clutch. Stieglitz and Wilson (1968) reported that the average clutch of 117 Florida duck nests was 9.4 eggs, and the range was 5 to 13. Clutch sizes decreased through the breeding season, with early nests averaging 10.1 eggs and later ones 8.9 eggs. The modal number of eggs in completed clutches was 10. Singleton (1953) reported that 108 nests of *maculosa* in Texas averaged 10.4 eggs per clutch. Eggs are apparently laid at the rate of one a day (Stutzenbaker, 1988).

Renesting is apparently prevalent, at least in the Texas mottled duck population. Engeling (1949) reported

Florida duck, adult pair

a case in which one female made five nesting attempts, laying a total of 34 eggs, before finally successfully hatching a brood of nine ducklings. Finger et al. (2003) reported an average clutch of 8.6 eggs in 26 initial Texas nests, 9.6 eggs in 10 second nestings, and 7.5 in 2 third-time efforts. Renesting has also been reported in Louisiana (Baker, 1983).

Incubation period. In a Florida study two wild nests hatched after 25 to 26 days of incubation, and two clutches that were hatched in an incubator had an incubation period of 26 days. From 21 to 30 hours elapsed between initial pipping and the hatching of the last egg (Stieglitz and Wilson, 1968).

Fledging period. Engeling (1949) noted that by six weeks of age young mottled ducks were fully feathered except for their wing feathers. Stutzenbaker (1988) reported fledging to occur at 63 to 70 days of age, but fledglings could make limited escape flights at 45 to 56 days.

Nest and egg losses. In a Florida study, 76.7 percent of 90 island nests hatched (Stieglitz and Wilson, 1968) with an average of 9 ducklings hatching from successful nests. However, in a Texas mottled duck study, only 10 of 46 nests were known to hatch. Of the remainder, predators destroyed 20, 9 were deserted, 5 were flooded, cattle trampled 1, and the fate of 2 was unknown. Direct or indirect destruction by dogs was the major source of predation in this study (Engeling, 1949), whereas in a Florida study avian predators, probably crows, destroyed 6 nests, but the nesting success rate was 77 percent for 117 nests (Stieglitz and Wilson, 1968). In another Texas study, a 96.2 percent hatching success and a 28 percent nesting success (108 nests) were reported (Singleton, 1953). Stutzenbaker (1988) reported a nesting success of 24.7 percent for 146 Texas nests.

Juvenile and adult mortality. Engeling (1949) estimated that an average brood of 8 or 8–9 mottled duck ducklings at hatching is normally reduced to 5 or 6 young at the time of fledging. Annual survival rates in Florida have been estimated at 50 percent for adult females (69 recoveries), 47.4 percent for immature females (145 recoveries), 54.8 percent for adult males (187 recoveries), and 90.9 percent for immature males (238 recoveries) (F. A. Johnson, cited by Moorman and Gray, 1994). Finger et al. (2003) estimated from radiomarked Texas females that brood survival during a wet year of 13 females was 69 percent, and that duckling survival to 30 days of age was 41 percent. However, during a dry year no broods survived.

General Ecology

Food and foraging. The only detailed study of food consumption is that of Beckwith and Hosford (1955), who analyzed the food contents of nearly 150 birds collected in all seasons in southern Florida. The yearly average for food intake was 87 percent vegetable origin. The highest incidence of consumption of animal material was from summer samples, when almost 40 percent, mostly water beetles, was of animal origin. Panic grass (*Panicum*) was the most important summer plant food, with smartweeds (*Persicaria*) in second place. Fall foods included seeds of ragweed (*Ambrosia*), paspalum, bristle grass (*Setaria*), panic grass, and smartweeds. Winter foods included spike sedge (*Eleocharis*), beak rush (*Rynchospora*), bulrush (*Scirpus*), fanwort (*Cabomba*), and ragweed. Major spring foods were smartweeds, cockspur (*Echinochloa*), bristle grass, and wax myrtle (*Cerothamnus*).

Sociality, densities, territoriality. These mallard-like taxa apparently do not differ greatly from northern mallards or black ducks in their social behaviors. Engeling (1950) noted that the "territory," more probably a home range, of one Texas *maculosa* pair was 0.5 mile in diameter. There is a record of a remarkably high nesting density on dredge-spoil islands at Indian River, Florida. The largest number of nests found on a single island was seven (Stieglitz and Wilson, 1968). Stieglitz and Wilson did not detect any territorial defense behavior in the dense nesting population of *fulvigula* that they studied. Three nests were once found in a 15-foot-diameter circle, two of which were within five feet of each other. At that study area there was an apparent nesting success rate of 77 percent for 117 nests; the island location probably provided some protection from some nest predators.

Interspecific relationships. As to major predators, at least in Florida and Texas, alligators are serious predators, especially during the molting period of late summer, and with droughts, which cause concentrations of both the ducks and alligators. The usual assortment of mammals, birds, reptiles, and other animals are no doubt also present; in a study by Johnson et al. (1995) the major known mortality factors for Florida-banded mottled ducks in descending importance were alligators, mammals, raptors, and vehicles.

The degree of social interactions among these southern populations with the mallard and black duck are still incompletely known but mostly occur during winter and are apparently limited. However, substantial interactions in the form of hybridization with northern mallards has been found in wild populations of all three taxa (see the Taxonomic Comments section). Quite possibly the relatively continuous pair-bonds that seem to be present in the southern mallard populations—paired Florida ducks have been seen every month—prevent more frequent mixed pairing, and early fall pairing behavior by resident mottled ducks occurs well before the mallards arrive and begin their courtship (Paulus, 1988).

General activity patterns and movements. The small amount of information so far available on these southern populations indicates that they are relatively sedentary. Engeling (1951) reported on 40 returns from mottled ducks banded in coastal Texas. None of the returns was from south of Aransas County, suggesting little or no southward movement during winter. The maximum movement was one of about 100 miles to the northeast. Hyde (1958) similarly noted that of 13 recoveries of birds banded in Florida the distance of movement ranged from 0 to 130 miles and averaged only 45 miles.

Social and Sexual Behavior

Flocking behavior. Apparently flock sizes in southern mallards are not normally very large. Beckwith and Hosford (1957) noted that in Florida duck flocks of up to 50 birds can be seen in August. By November they are usually in groups of 6 to 20 birds. Flocks of up to 13 birds are seen until late February, when the birds break up into units of pairs, trios, and single birds. Aldrich and Baer (1970) reported that a wintering flock of at least 1,000 *diazi* was seen during January in Mexico, but counts made during May resulted in a total count of 120 ducks on 14 different areas, or fewer than 10 birds per observation site.

Pair-forming behavior. Relatively few observations on pair-forming behavior have been seen in wild birds, which is a further indication that pair-bonds may be relatively continuous under natural conditions. Among captive specimens of *fulvigula* the normal mallard repertoire of social displays has been observed (Johnsgard, 1959, 1965; see Fig. 18), and the same appears to be true of wild *maculosa* (Weeks, 1969; Paulus, 1980) as well as of *diazi* (Williams, 1980).

Copulatory behavior. Copulatory behavior takes the same form in these southern populations as is typical of northern mallards and American black ducks (Johnsgard, 1965; Weeks, 1969; Paulus, 1980).

Nesting and brooding behavior. During the incubation period, the female probably normally leaves the nest once or twice a day, for periods of about two hours. The time at which the male deserts his female to begin the postnuptial molt probably varies considerably, but Engeling (1950) believed that male mottled ducks might remain with their mates until about the time of hatching. The young remain in the nest from 12 to 24 hours, and the female leads them away from the area of the nest 24 to 48 hours after hatching (Stieglitz and Wilson, 1968).

Postbreeding behavior. In Texas, the birds move from open prairie areas to rice fields and marshes after breeding (Engeling, 1951). The postnuptial molt involving a several-week flightless period is present (Beckwith and Hosford, 1957), which in Louisiana mottled ducks lasts 27 days (Paulus, 1984). Courtship there begins as early as August, and copulations occur as early as September, by which time 70 percent of the females may already be paired (Paulus, 1988).

American Black Duck
Anas (p.) rubripes Brewster 1902

Note: The parenthetic *(p.)* indicates that this taxon is here considered to be part of the common mallard super-species *Anas platyrhynchos*.

Other vernacular names. Black mallard, red-legged black duck

Range. Breeds from Manitoba and Ontario eastward to Labrador and Newfoundland, south to Minnesota, and east through the Great Lakes states to the Atlantic coast as far south as North Carolina. Winters through the eastern United States, mostly coastally, from southern Canada and Maine south to Georgia, and increasingly rarely to Florida and the Gulf coast.

Subspecies. None recognized. However, based on the long-term and still increasing incidence of introgressive hybridization in eastern North America, *rubripes* might better be recognized as a subspecies of *platyrhynchos* (Johnsgard, 1959, 1961c), in which case the vernacular name "black mallard" would be most appropriate. Ankney et al. (1986), using mean genetic distance measures, stated that "our genetic data do not support even subspecific status for the black duck," and noted that there is as much genetic variation within the black duck's population as there is between common mallards and black ducks. I have here designated the American black duck as being part of the common mallard's superspecies, the black duck being thus classified as *A. (p.) rubripes*, and, along with the mottled duck, Florida duck, and some other insular relatives, recognized as genetically fitting within the common mallard's potentially panmictic gene pool.

Measurements. *Folded wing:* Delacour (1954): Males 265–292 mm; females 245–275 mm. Kear (2005): 377 males, average 285.0 mm; 355 females, average 268.7 mm.

Culmen (bill): Delacour (1954): Males 52–58 mm; females 45–53 mm. Kear (2005): 377 males, average 54.3 mm; 355 females, average 51.1 mm.

Weights (mass). Nelson and Martin (1953): 366 males, average 2.7 lb. (1,224 g); 297 females, average 2.4 lb. (1,088 g). Jahn and Hunt (1964): 86 adult males, average 2.94 lb. (1,332 g); 185 immature males, average 2.69 lb. (1,219 g); 80 adult females, average 2.56 lb. (1,162 g); 172 immature females, average 2.44 lb. (1,106 g). Kear (2005): 222 adult males (fall), average 1,317 g; 355 adult females (fall), average 1,090 g.

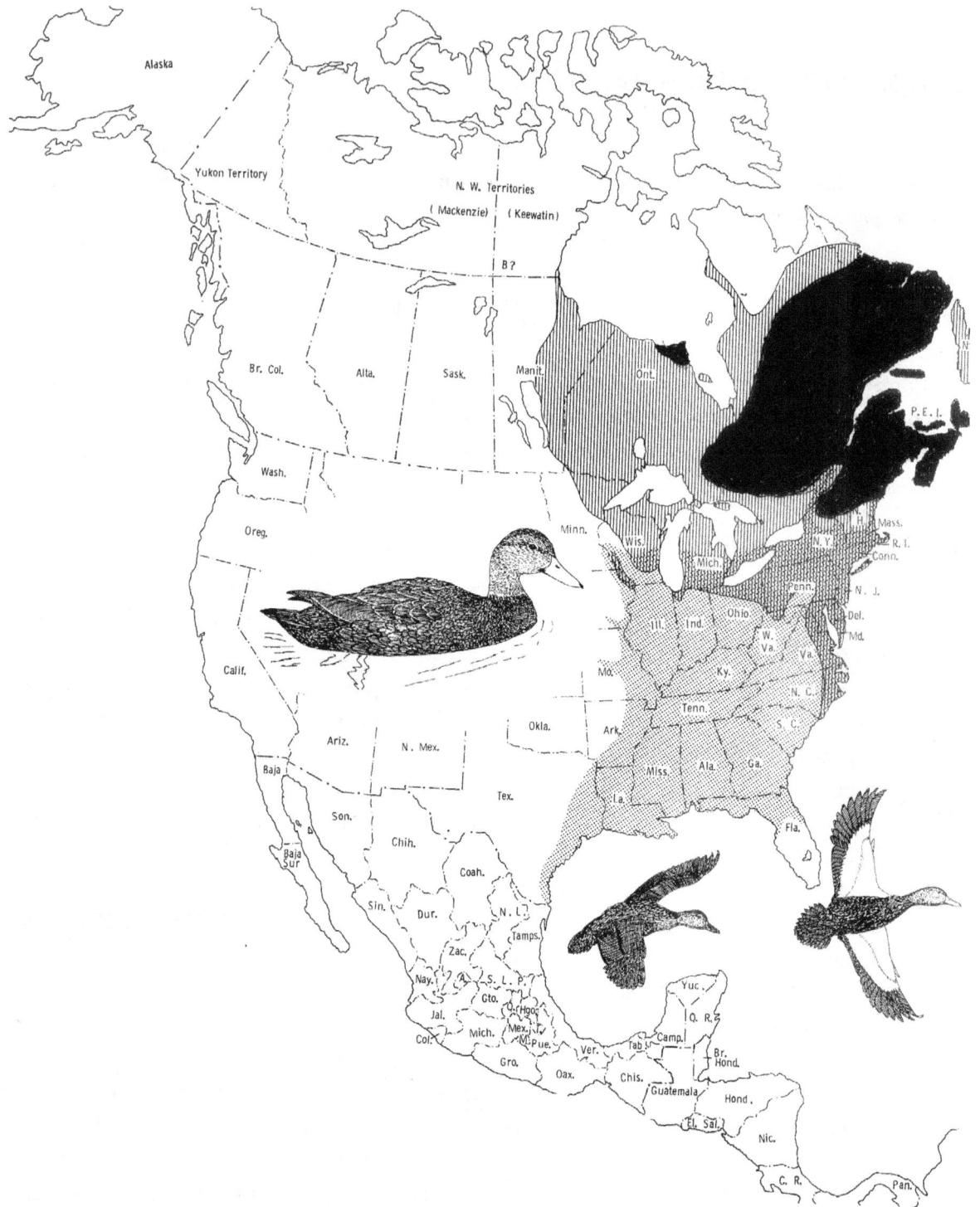

The breeding (vertical hatching, with denser concentrations inked) and wintering (shaded) range of the American black duck.

Identification

In the hand. Black ducks may be readily identified in the hand by their mallard-like shape and size, an almost entirely brownish black body color, and the absence of any white anterior to the speculum. Little or no white is normally present on the trailing edge of the secondaries, but hybridization with mallards has increasingly diluted the purity of most black duck populations, so this criterion is not absolute. Female mallard–black duck hybrids most resemble mottled ducks but usually show some white on the greater secondary coverts, especially on the outer web (Johnsgard, 1959). Male hybrids usually show some green iridescence behind the eyes, often forming a fairly distinctive green patch.

In the field. The dark body with only slightly lighter head color makes black ducks conspicuous in any gathering of ducks. They are mallard-like in every respect except their coloration, including their vocalizations. In flight, the white under wing-coverts contrast more strongly with the dark body and upper wing coloration than is true of mallards, and this flashing wing pattern of dusky and white makes black ducks recognizable for as far away as they can be seen. When in breeding condition, the brilliant yellow bill of the male is very conspicuous and allows for ready sexual identification.

Age and Sex Criteria

Sex identification. External features that identify a male are a bright yellow bill that lacks spotting, breast feathers with rounded light markings centrally (instead of V-shaped markings) or no light central markings at all, and bright reddish rather than brownish feet. Cloacal or internal examination is the most reliable sexing method for younger birds.

Age determination. Immature birds are likely to have small, frayed, or faded tertials and tertial coverts compared to larger and freshly grown feathers in adults. The middle coverts of immatures may be narrow and somewhat trapezoidal, especially just anterior to the tertial coverts (Carney, 1964). Immatures might also exhibit notched tail feathers, especially during their first fall.

Distribution and Habitat

Breeding distribution and habitat. To a degree greater than any other North American waterfowl species, the black duck is largely limited to the eastern, forested portion of the continent. In Canada its summer range extends westward only to eastern Manitoba, where it is generally scarce (Godfrey, 1966). In Saskatchewan there are a few scattered records (Murray, 1959), and also a few from Alberta, where it has been reported to nest (Godfrey, 1966). From Ontario eastward to Newfoundland it is the commonest breeding duck species in most areas, at least as far north as the tree line.

In the United States the black duck is largely a breeding bird of the eastern forests and coastal marshes, as Stewart (1958) has pointed out. He listed two areas of high breeding population densities, the hemlock–white pine northern hardwood forest region east of longitude 85°W, and the tidewater areas of Delaware Bay and the eastern shore of Chesapeake Bay, Maryland. Boreal coniferous forests and tidewater areas to the north of Maryland support medium breeding densities, while low breeding densities occur in tidewater areas south to North Carolina and in several forest associations. These are the boreal coniferous and hemlock–white pine–northern hardwood regions west of longitude 85°W, the maple-basswood forest region, and northern parts of the beech-maple, mixed mesophytic, and oak-chestnut forest regions as defined by Braun (1950).

Although Minnesota represents the normal western limit of black duck breeding habitat in the United States, there have been a few isolated records of nesting in North Dakota. In spite of the regular occurrence of black ducks in hunter kills along the Central Flyway states from North Dakota to Oklahoma, there is no indication that the black duck is now significantly extending its breeding range to the west (Johnsgard, 1961b).

Stewart stated that typical interior breeding habitats include alkaline marshes, acid bogs and muskegs, lakes and ponds, and the margins of streams, while in tidewater areas black ducks breed in salt, brackish, and fresh marshes as well as in the margins of bays and estuaries. Stotts and Davis (1960) noted that of 731 nests found, almost 60 percent were in wooded habitats, versus 17 percent in marshes.

Population. Current (2016) evidence indicates that the black duck has been in a slow, long-term population decline in eastern North America, especially relative to mallards in the same region. The 2013 national breeding population estimate for black ducks was 600,000, or close to the 20-year 1990–2012 average, compared with a decline of more than 50 percent during the three decades from the 1950s to the 1980s. Hunter-kill estimates of black ducks in the Atlantic Flyway during the early 2000s have dropped to nearly one-third of those typical of the late 1960s (the five-year average from 2009 to 2013 was 77,000). In the Atlantic Flyway mallard kills averaged 311,000 birds during the 2014–15 seasons, and black ducks 65,000, suggesting that the black duck now (2016) represents about 17 percent of the combined mallard/black duck gene pool in the Atlantic Flyway. By comparison, during the ten-year period 1949–58 the average total numbers of mallards in the Atlantic Flyway was 218,000, and that of the black duck was 457,000, so black ducks then composed 68 percent of their combined population (Johnsgard, 1961c).

During that same period mallard populations have thrived. The 2013 breeding survey produced a national estimate of 10.3 million mallards, a 17:1 population ratio favoring mallards over black ducks (US-FWS, 2014). Additionally, the average hunter kill in the United States averaged 3.77 million mallards shot during the 2014–15 seasons. The corresponding average black duck kills were 86,000 for these two seasons (Raftovich, Chandler, and Wilkins, 2015). This contrasts with the long-term average of about 8,000 hybrids taken in the Atlantic Flyway during the early 2000s, or nearly 10 percent of total recent average kills for the black duck in that flyway. This estimate of hybrid frequencies is 3.7 times higher than those I calculated for the 1960s, when I estimated an incidence of 2.7 percent hybrids relative to the black duck sample in the Atlantic Flyway (Johnsgard, 1961c, 1967).

Fig. 20. Male American black duck.

Since then, black duck populations in the Mississippi Flyway have apparently declined to an even greater degree than has occurred in the Atlantic Flyway, both relative to mallards and in terms of actual population size. Based on USFWS Midwinter Surveys, black ducks totaled 223,000 in 2010, which was 7 percent lower than the 2000–09 average. Of the total 223,000 birds, the Atlantic Flyway component was 5 percent below the ten-year flyway average, while the Mississippi Flyway portion was 25 percent below that flyway's average. More telling was the fact that Atlantic Flyway black ducks declined 55 percent between the 1955–59 Midwinter Survey average and the 2000–05 average (from 403,000 to 221,000), while during that same period Mississippi Flyway black ducks declined by 86 percent (from 213,000 to 30,000).

During the Atlantic Flyway Breeding Waterfowl Plot Surveys, black ducks totaled 38,200 birds, a 43 percent decline from the 1993–2009 average. Meanwhile, mallard population estimates in North America were then about 13 million birds, including a long-term (1990–2010) average of 764,000 in the Atlantic Flyway (Baldassarre, 2014), or 30 times greater than the Atlantic Flyway black ducks.

The estimated total annual Canadian kill of black ducks in 2014 was 63,000, compared with a mallard kill of 500,000. The respective 2014 US kills were 72,300 black ducks and 3.9 million mallards. During the 2014–15 US hunting seasons, about 4,200 birds identified as black duck–mallard hybrids were also shot.

With a US hunter kill of black ducks of about 72,000 in 2014, the hybrids constituted about 5 percent of the collective black duck–hybrid kill. Like the southern mallards, the American black duck's genome is being increasingly threatened by declining black duck numbers, increasing genetic dilution by the far more abundant common mallard over the long term. Brodsky and Weatherhead (1984) and Brodsky, Weatherhead, and Dennis (1988) have pointed out the behavioral superiority of male mallards when competing with black ducks for mates, which places the black duck's future in further jeopardy The black duck's status as a species-level taxon is thus becoming progressively more questionable, as I first predicted in 1959, and which has been later supported with genetic studies by Ankney et al. (1986); Ankney, Dennis, and Bailey (1987); and Mank, Carlson, and Brittingham (2004).

Wintering distribution and habitat. During the mid-1900s wintering black ducks could be found over a wide geographic range from Minnesota and coastal Texas on the west to the Atlantic coast from northern Florida to Nova Scotia on the east (Johnsgard, 1959). Stewart (1958) indicated that wintering black ducks were characteristically found within the eastern deciduous forest formation and tended to concentrate on coastal tidewaters and on the larger streams, lakes, and reservoirs of the interior. The heaviest coastal concentrations occurred from North Carolina to Massachusetts, but large numbers also occurred on the rivers of Tennessee, Kentucky, Ohio, Indiana, and Illinois (Johnsgard, 1959). Geis et al. (1971) noted a similar pattern of wintering concentrations and also indicated that the western end of Lake Erie and the Atlantic coastline north to Nova Scotia were areas of winter concentrations.

Stewart (1962) noted that migrant and wintering black ducks in the Chesapeake Bay area occupied a greater variety of habitats than any other waterfowl species, but brackish estuary bays with extensive adjacent agricultural lands were strongly favored. Estuarine bay marshes, especially those with salt water, also received high usage, as did coastal salt marshes and adjacent impoundments. In general, black ducks showed a higher usage of saltwater habitats than did mallards, which concentrated on fresh to brackish water areas.

Since that period, black ducks have increasingly concentrated in the Atlantic Flyway, with Midwinter Surveys indicating that nearly 90 percent have more recently been found there, with the Mississippi Flyway accounting for nearly all of the remainder. Coastal areas from Maine to Georgia are heavily used, especially those between New York and North Carolina. In particular, Chesapeake Bay and the coastal wetlands of New Jersey apparently account for more than half of the entire wintering Atlantic coast population, with most of the rest wintering from Long Island northward. Although fairly small numbers winter in Canada except for southern Ontario, global warming has allowed a major northward winter shift of water-dependent birds in the Great Plains since the 1960s (Johnsgard, 2015), and the same trend can be detected in Ontario (Brook et al., 2009).

General Biology

Age at maturity. Like mallards, black ducks are known to be sexually mature their first year. Coulter and Miller (1968) found that first-year female black ducks had clutch sizes that were below the average they

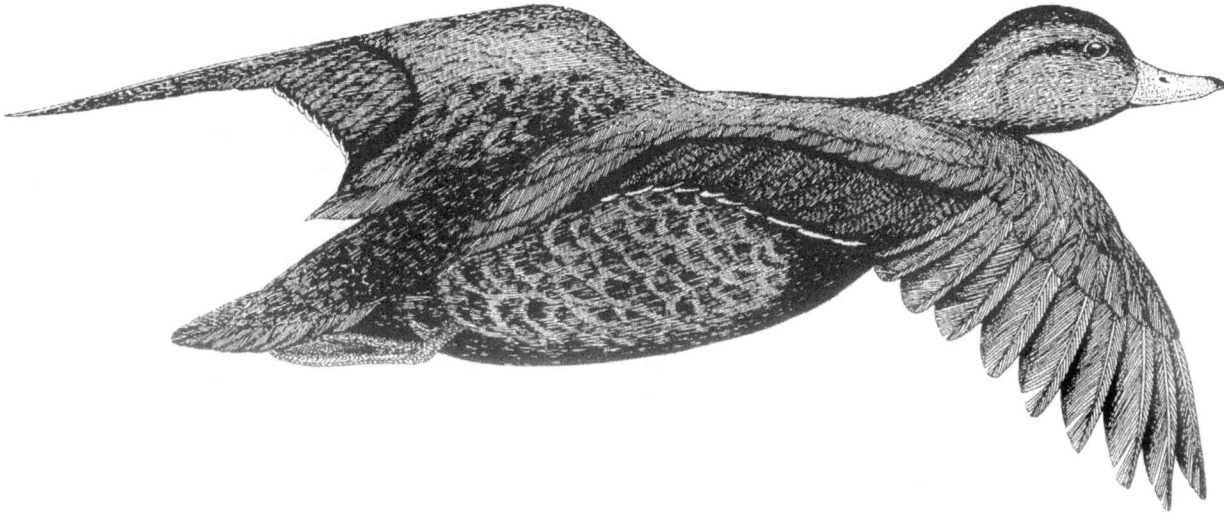

Fig. 21. Male American black duck, in flight.

found for the species (8.4 versus 9.5 eggs) and that only one of seven yearling hens renested after their nests were removed.

Pair-bond pattern. Pair-bonds are broken during the incubation period and are reestablished in the fall during social courtship. The incidence of older adults re-pairing with their earlier mates seems to be fairly low (Stotts, 1968).

Nest location. Stotts (1955) reported that of 356 nests found in the Kent Island area of Maryland, about 80 percent were near the margins of wooded areas, with marshes and cultivated fields being second in frequency of usage. The study by Stotts and David (1960) indicated that honeysuckle (*Lonicera*) and poison ivy (*Rhus radicans*) were favored covers, accounting for 43.3 percent of 593 nests, whereas brush or tree cover accounted for 32.1 percent and marsh grasses 14.0 percent.

Coulter and Miller (1968) noted that among nests found in sedge-meadow bogs over a 14-year period at Goose River, Maine, more than 80 percent were associated with leatherleaf (*Chamaedaphne*) and sweet gale (*Myrica*) as principal cover plants. Leatherleaf's preferential use for overhead cover is apparently related to its characteristic low-growing, densely branched, growth form and its nearly persistent leaves. Additionally, its extensive roots form small hummocks that are elevated above the damp floor of the marsh, making it an excellent nest site. On wooded islands, cover usage was quite different, with sites being selected that offered the best concealment in places where ground litter was also available. These sites included live conifers, blueberry (*Vaccinium*) bushes, dead or fallen woody growth, and live or dead herbaceous plants, especially nettle (*Urtica*). Coulter and Miller found that island nesting by black ducks was common only where sedge-meadow covers or other marsh nesting covers are not available.

Clutch size. Clutch sizes reported in the literature generally range from 9.1 to 9.5 eggs, with the former reported by Stotts and Davis (1960) and the latter by Coulter and Miller (1968). Coulter and Miller's study, based on 620 clutches, indicated a range in clutch sizes from 4 to 15 eggs, with nearly 50 percent of the clutches having either 9 or 10 eggs. They found a decrease in average clutch size as the season progressed, a larger average clutch size produced by females known to be at least two years old compared with birds of mixed ages, and a slightly larger average clutch size for first nests over renests by the same birds. In two of 22 cases the rate of egg-laying deviated from one per day, but disturbance may have caused these variations.

Incubation period. The incubation period of black ducks is very similar to that of mallards, about 27 days. Stotts and Davis (1960) estimated the average incubation period to be 26.2 days, with a range of 23 to 33 days. The incubation periods were slightly shorter in artificially incubated eggs than in the naturally incubated.

Fledging period. The black duck fledging period was reported as 7.5 weeks (52–53 days) by Wright (1954) and as 8.5 weeks (59–60 days) by Lee et al. (1964a). It is evidently very similar to that of mallards (52–60 days).

Nest and egg losses. In a study by Stotts and Davis (1960), only 38 percent of 574 nests were terminated by hatching one or more eggs, and 15 percent of the eggs in successful nests did not hatch. Fully half of the nests studied were destroyed by predators, 34 percent by crows alone, while raccoons also destroyed a considerable number. Besides destroying whole clutches, crows (mostly fish crows) also removed almost 10 percent of the eggs from nests that later were successfully terminated. Wright (1954) estimated that an average of eight eggs are normally hatched per successful nest during his studies in Canada. Summarizing various studies, Jahn and Hunt (1964) judged that an average of 64 percent of the females succeeded in hatching broods.

Coulter and Miller (1968) estimated a 31 percent renesting rate in black ducks, compared with an earlier estimate of 16 percent by Stotts and Davis. The former authors reported a surprisingly high (77 percent) hatching success in renesting attempts but did not indicate the hatching success of initial nesting attempts. Other studies (Coulter and Miller, 1968; Dwyer and Baldassarre, 1993) have indicated that mallards are more persistent renesters than are black ducks.

Juvenile mortality. Wright (1954) estimated that black duck broods average about 8 ducklings for broods under two weeks of age and that an additional average of 1.7 ducklings are lost during the first six weeks of life, so about 6 ducklings per successful brood may be expected to fledge. Jahn and Hunt (1964) summarized several studies and estimated that 6.9 young per female are reared to fledging. Later mortality rates of juvenile birds are substantially higher than those of adults; Geis et al. (1971) estimated a 64.9 percent first-year mortality rate for birds banded as immatures.

American black duck, adult pair

Adult mortality. Geis et al. (1971) estimated that the annual adult mortality rate for banded black ducks was about 40 percent (60 percent survival rate) for adults of both sexes, with females having a considerably higher mortality rate than males. Thus adult males had an approximate 62 percent annual survival rate, compared to 53 percent for females.

Francis, Sauer, and Serie (1998) estimated black duck survival rates over six regions and three time periods from 1950 to 1993. The survival rates for the entire approximate four-decade period were 66.1 percent for adult males, 58.7 percent for adult females, 56.3 percent for immature males, and 53.3 percent for immature females. Krementz et al. (1989) also calculated survival rates from 1950 to 1983; annual survival rates were 61.5 percent for adult males, 45.1 percent for adult females, 44.3 percent for immature males, and 35 percent for immature females. From 32 to 47.7 percent of the estimated overall mortality was estimated to be the result of hunting.

General Ecology

Food and foraging. Perhaps because it tends to inhabit more distinctly salty water on its coastal wintering grounds, the black duck consumes a higher proportion of food of animal origin than does the mallard. In coastal bays about half the total food intake may be of mollusks, especially univalve mollusks (Martin et al., 1951). However, even in brackish estuaries the black duck sometimes feeds heavily on the leaves, stems, and rootstalks of submerged aquatic plants, the seeds of submerged and emergent plants, and the rootstalks of emergent marsh plants (Stewart, 1962). Stewart found the univalve *Melampus* commonly represented in birds taken in salt or brackish water; the bivalve *Macoma* was found in somewhat fewer samples.

Hartman's (1963) study of fall and winter foods of black ducks shot on the Penobscot estuary, Maine, has emphasized the importance of *Macoma* and *Mya* clams as food of this species; these two genera of mollusks accounted for nearly half of the identified food materials by volume. Important plant foods included acorns, the stems and leaves of cordgrass (*Spartina*), and the seeds of various sedges (*Carex*) and bulrushes (*Scirpus*). Mendall's (1949) study of Maine black duck foods showed a similar high incidence of mollusk consumption during winter, while foods taken at other seasons were predominantly of vegetable origin.

Although the black duck obtains most of its food from the surface or from what it can reach by tipping-up, it has been known on several occasions to dive for food (Kear and Johnsgard, 1968). Likewise, field-feeding in grain fields is almost as common among black ducks as among mallards, at least where both species occur together. Winner (1959) described the field-feeding periodicities of both the mallard and black duck in Ohio and found that mixed foraging flocks of the two species were prevalent.

Sociality, densities, territoriality. Like the mallard, black ducks congregate in extremely large numbers during fall and winter wherever the combination of open water and sufficient food supplies can be found. By spring, the flock sizes begin to decrease as paired birds start to avoid unpaired males.

Although Stotts (1957) reported some unusually high nesting densities on certain islands of Chesapeake Bay (up to 21.4 nests per acre), these were clearly artifacts of island nesting. Coulter and Miller (1968) also reported maximum densities of about five nests per acre on an island in Lake Champlain. However, in the preferred bog-nesting habitats of Maine, densities were never higher than one nest per 20 to 40 acres, and similarly Stewart (1962) found a breeding density of a pair per 19 acres on a 1,000-acre area of brackish estuarine bay marsh in Maryland. Jahn and Hunt (1964) reported similar breeding densities in Wisconsin. Thus a nesting density of about one pair per 20 acres would seem typical of high-quality, non-island breeding habitat.

Divergent opinions as to the existence of territorial behavior in black ducks have appeared in the literature (Stotts and Davis, 1960), and the evidence favoring such behavior in this species is not convincing. Stotts and Davis described several instances of aggression, which they attributed to territoriality, but noted that it was most evident in late April and May, when most renesting was in progress. This would clearly indicate that typical territoriality was not involved and that aggressive or sexual behavior associated with attempted renesting was responsible for much of the apparent territoriality.

American black duck, female and brood

Interspecific relationships. The close evolutionary relationships between black ducks and mallards have been previously studied (Johnsgard, 1959, 1961c), and a significant rate of natural hybridization since then has been established. This interaction has risen sharply in recent years, as mallards have moved increasingly eastward as wintering and breeding birds. In one early study (Goodwin, 1956) it was found that, in spite of fairly frequent hybridization, mallards increased rapidly in proportion to black ducks in the combined population. This may be brought about by nonselective mating or by tendencies toward cross-matings in the case of female black ducks, which tend to favor mating with mallards. On the other hand, ecological differences in the form of habitat breeding preferences tend to keep the two forms separated on their breeding grounds and probably militate against the maintenance of mixed pairings (Johnsgard, 1959, 1967a). The primary zone of contact between mallards and black ducks has moved considerably eastward during the past century, and current evidence indicates that hybridization between them will continue to increase (see Breeding Distribution and Habitat section).

General activity patterns and movements. Winner's (1959) study on the field-feeding behavior of mallards and black ducks indicated that mallards tend to leave for the evening feeding flight earlier than black ducks, although mixed flocks were often seen. Field-feeding behavior by black ducks may be relatively less common than in mallards; Mendall (1949) found that only a small proportion of black ducks in Maine's grain-growing district actually consume grain, and noted that crop damage by black ducks is very rare. Little preference is shown there among black ducks for fields containing oats, buckwheat, or barley. However, development of a grain-feeding "tradition" among black ducks may become increasingly likely as mallards become more abundant in the eastern states and mixed flocks become more frequent.

Social and Sexual Behavior

Flocking behavior. Black ducks are seemingly almost identical to mallards in their flocking behavior, congregating during fall and winter wherever the combination of water and safe foraging areas exists, sometimes massing in flocks of several thousand birds. In spite of the flock size, the basic unit composition is that of individual pairs of birds and generally small groups of unpaired males and females. As the percentage of obviously paired birds increases during the winter, the flock sizes tend to decrease.

Pair-forming behavior. Pair-forming behavior in black ducks has a seasonal pattern very similar to that of mallards. Adult birds that had been previously paired and meet again after molting probably re-pair without any ceremony, thus accounting for the low percentage of paired birds seen in August (Stotts, 1958). Other adults begin social display in September or October, but it is probable that immature females do not begin pair-forming activity until they are six or seven months old, and young males when slightly older (Stotts and Davis, 1960). This would account for the sharp increase in apparently paired birds seen between November and January (Johnsgard, 1960b). The highest incidence of apparently paired birds is in April, when virtually all females appear to be paired. Although Stotts (1958) noted a maximum pair incidence of about 90 percent, the excess of males in wild populations prevents some males from obtaining mates.

Actual pair-forming mechanisms, as well as the motor patterns and vocalizations associated with social display, appear to be virtually identical in mallards and black ducks (Johnsgard, 1960b). Mixed courting groups frequently occur in areas where the two species have overlapping ranges, and mixed pairs involving both of the two possible pairing combinations have been seen.

Copulatory behavior. Precopulatory and postcopulatory behavior patterns of black ducks are identical to those of mallards (Johnsgard, 1965).

Nesting and brooding behavior. Females deposit eggs in the nest at the rate of about one per day, with most egg-laying occurring fairly early in the morning and often within two hours after sunrise. Males rarely accompany their mates to the nest during egg laying but rather typically wait at a customary loafing site that is often the point of water nearest the nest. A down lining usually begins to appear when the clutch is about half complete and typically becomes profuse just before incubation begins. Unlike their behavior early in incubation, females rarely leave their nests during the last few days prior to hatching. Pipping usually takes about 24 to 30 hours from the time cracks first appear on the egg, and at that time the female typically begins to perform "broken-wing" behavior if disturbed on the nest (Stotts and Davis, 1960). Stotts and Davis also determined that the average attendance period of males with females following the start of incubation was 14.3 days, with a range of 7 to 22. In the case of renesting females, the average period of male attendance was 9.1 days. Thus, in many cases, the original mate was present long enough to fertilize the female for an attempted renest.

Postbreeding behavior. Following the male's desertion of his mate, he begins to undergo his postnuptial molt and enters a flightless period that probably lasts about four weeks. At this time the birds are usually wary and are rarely seen. There is no clear evidence of any substantial molt migration of male black ducks to specific molting areas. However, Hochbaum (1944) mentioned that a few male black ducks molt in the Delta, Manitoba, marshes, and the birds summering near Churchill, Manitoba, may also be mostly postbreeding males (Godfrey, 1966). Likewise, the female deserts her brood at about the time they become fledged, or at some stage prior to this time, and also begins her postnuptial molt. By August both sexes are again flying and starting to gather with immature birds in favored foraging areas.

White-cheeked Pintail
Anas bahamensis Linnaeus 1758

Other vernacular names. Bahama duck, Bahama teal

Range. The Bahamas islands, the West Indies, Colombia, eastern South America from Venezuela and northeastern Brazil to central Argentina, and west of the Andes from Ecuador to central Chile, plus the Galapagos Islands, with rare stragglers reaching the southeastern United States.

North American subspecies. *Anas b. bahamensis* L.: Lesser White-cheeked Pintail. The Bahamas islands, the West Indies (Greater and Lesser Antilles), and northern South America.

Measurements. *Folded wing:* Delacour (1956, race not specified): Males 211–217 mm; females 201–207 mm. Kear (2005): Males of *A. b. bahamensis* 201–231 mm, average of 68, 220 mm; females 180–220 mm, average of 50, 207 mm.
   Culmen (bill): Delacour (1956, race not specified): Males 42–44 mm; females 40–43 mm. Kear (2005): Males of *A. b. bahamensis* 34–49 mm, average of 68, 45 mm; females 39–47 mm, average of 50, 42 mm.

Weights (mass). Weller (1968): 7 adult males of *A. b. rubrirostris*, average 710.4 g (1.57 lb.); 4 adult females, average 670.5 g (1.48 lb.). Haverschmidt (1968): males of *A. b. bahamensis* 474–533 g; females 505–633 g. Kear (2005): 68 males of *A. b. bahamensis* 440–630g, average 526 g; 50 females 395–650 g, average 502 g.

Identification

In the hand. This dabbling duck could only be easily confused with the far more common northern pintail, since both have elongated central tail feathers. However, the white-cheeked pintail's central feathers are of the same reddish buff color as the more lateral tail feathers, and no other North American species of duck has white cheeks and a throat that sharply contrast with a uniformly dark brown on the rest of the head. Likewise, the red marks at the base of the bluish bill are unique.

In the field. The field marks for this rare but distinctive species are simple: a generally reddish brown duck with white extending from the cheeks to the base of the neck, red spots on the side of the bill, and a pointed tail. It is considerably smaller than the northern pintail but has the same general body profile. In flight, it also exhibits a similar pattern of white, gray, and dark brown on the under wing-coverts but is otherwise much more reddish buff than the northern pintail. The male utters a weak *geeee* sound during courtship display, and the female's calls are scarcely distinct from those of the northern pintail.

134

Fig. 22. Male white-cheeked pintail.

Age and Sex Criteria

Sex determination. Adult males have a distinctly more brilliant red color at the base of the bill and more immaculate white cheeks and throat than do the females. The tail is also longer (female maximum 84 mm, adult male minimum 98 mm).

Age determination. First-year birds no doubt exhibit notched tail feathers, and the tail is shorter and less pointed than in adults. The iris color is brown rather than red or brownish red. The pale orange bill coloration becomes a brighter red as sexual maturity approaches (Kear 2005).

White-cheeked pintail, adult pair

Occurrence in North America

In spite of the large number of recent records of this species in North America, there are very few old records. Bent (1923) listed only a single record for Florida in 1912, and there were historic Virginia and Wisconsin records. However, since the 1960s a remarkable number of sightings were made in a variety of Florida locations, including Pasco County, Fort Lauderdale, Lantana, West Palm Beach, Everglades National Park, and Loxahatchee National Wildlife Refuge. Beyond these Florida sightings, there were also sightings or specimen records from Alabama, Delaware, and Illinois by the mid-1970s.

A recent (2016) eBird map indicated sight records from many Florida locations, north on the Atlantic side to Scottsmoor, several from Merritt Island, Pelican Island National Wildlife Refuge, and Everglades Wildlife Management area, and one each from T. M. Goodwin Waterfowl Management Area and Grassy Waters Nature Preserve. On Florida's Gulf coast there are records north to Fillman Bayou and others southward from Tampa Bay and Myakka River State Park.

Farther north, there are several dozen eBird sightings from Chincoteague National Wildlife Refuge, Maryland, and sightings have been made at Ridgeway Park, near Newport News, and Back Bay National Wildlife Refuge, Virginia, There is also a Texas record from Laguna Atascosa National Wildlife Refuge. It is of course possible that some of these distant records represent escapes from captivity.

Northern Pintail
Anas acuta Linnaeus 1758

Other vernacular names. American pintail, common pintail, sprig, sprigtail

Range. Breeds through much of the Northern Hemisphere, in North America from Alaska south to California and east to the Great Lakes and eastern Canada, in Greenland, Iceland, Europe, and Asia as well as in the Kerguelen and the Crozet Islands. Winters in the southern parts of its breeding range in North America, south to Central America and northern South America.

North American subspecies. *Anas a. acuta* L.: Northern Pintail. Range as indicated above, except for the Kerguelen and the Crozet Islands.

Measurements. *Folded wing:* Delacour (1956): Males 254–287 mm; females 242–266 mm. Owen (1977): Adult males average 269.4 mm; females average 254.1 mm.

Culmen (bill): Delacour (1956): Males 48–59 mm; females 45–50 mm. Owen (1977): Adult males average 52.0 mm; females average 47.6 mm.

Weights (mass). Nelson and Martin (1953): 937 males, average 2.2 lb. (997 g), maximum 3.4 lb. (1,450 g); 498 females average 1.8 lb. (815 g), maximum 2.4 lb. (1,087 g). Bellrose and Hawkins (1947): 237 adult males, average 2.28 lb. (1,034 g); 403 immature males, average 2.15 lb. (975 g); 60 adult females, average 1. 96 lb. (888 g); 219 immature females, average 1.84 lb. (834 g). Owen (1977): Adult males average 915 g; females average 783 g.

Identification

In the hand. A northern pintail of either sex may be recognized in the hand by its slim-bodied and long-necked profile, sharply pointed rather than rounded tail, gray feet, gray to grayish blue bill, and a speculum that varies from brownish or bronze to coppery green, with a pale cinnamon anterior border and a white trailing edge. Another long-tailed species, the oldsquaw, has a large lobe on the hind toe, the outer toe as long or longer than the middle toe, and secondaries that lack iridescence or a white trailing edge.

In the field. The streamlined, sleek body profile of northern pintails is apparent on the water and in the air. When on the water, males exhibit more white than any other dabbling duck; their white breasts and necks can be seen for a half mile or more. When closer, the dark brown head, often appearing almost blackish, is

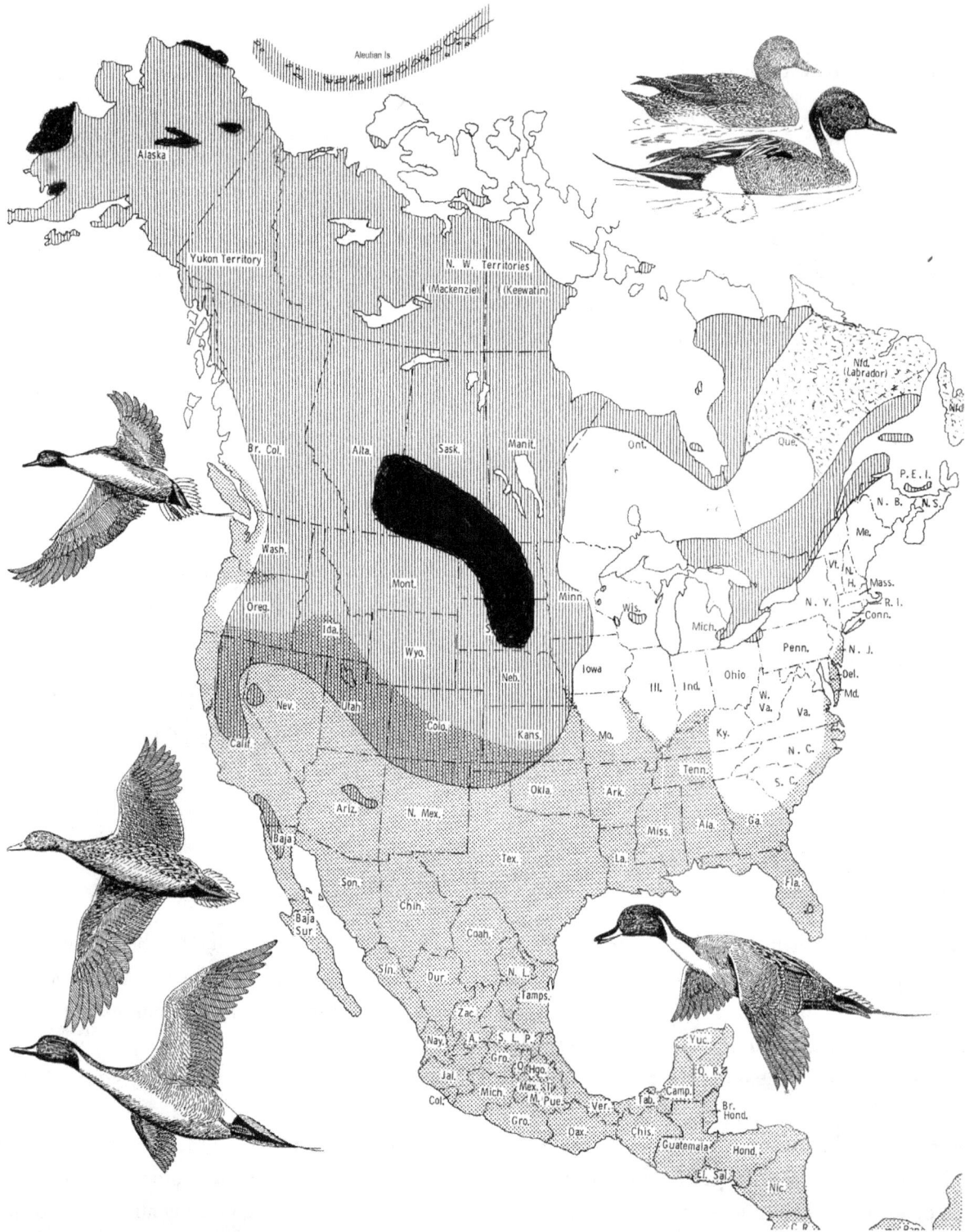

The breeding (vertical hatching, with denser concentrations inked), wintering (shaded), and acquired or marginal (stippled) range of the northern pintail.

apparent, as are the grayish flanks, separated from the black under tail-coverts by a white patch on the sides of the rump. Females are somewhat smaller, mostly brownish ducks, with a dark bill that shows no trace of yellow or orange, and they show no conspicuous dark eye-stripe or pale spot on the lores as in some other female dabbling ducks. During winter and early spring, males spend much time in courtship display, and one of their distinctive courtship calls, a fluty *pfüh*, can often be heard before the birds are seen either in flight or on the water. The quacking notes of female pintails are not as loud as those of female mallards, and the decrescendo series of notes is usually rather abbreviated.

Age and Sex Criteria

Sex determination. An iridescent bronzy speculum with a black bar in front of a white tip indicates a male, as does the presence of tertials that are long and gray with a wide black stripe. Vermiculations on the scapulars or elsewhere also indicate a male, but juvenile males may lack both vermiculations and the speculum characteristics mentioned above. Thus, juvenile birds may have to be examined internally to be certain of their sex (Carney, 1964).

Age determination. In males, the tertial coverts of juveniles are edged with a light yellowish brown, while those of older males are without such light edges. In immature males, the middle coverts are also similarly edged and may appear narrow, rough, and frayed. In females, the tertial coverts of immature birds are also narrow and frayed, and the middle coverts are narrow and somewhat trapezoidal with barring near the feathers' edges, rather than being rounded with barring between the edge and the feather shaft (Carney, 1964). The tail feathers may also have notched tips until they are molted in fall or winter. See also Esler and Grand (1994) for age determination in spring birds.

Distribution and Habitat

Breeding distribution and habitat. One of the most widely distributed of all North American ducks, the northern pintail breeds from the Aleutian Islands on the west to the Ungava Peninsula on the east, and from northern Texas and New Mexico on the south to at least as far north as Victoria Island, Northwest Territories. There is even a record of a brood at 82°N latitude on Ellesmere Island, some 700 miles north of the previously known breeding limits (Maher and Nettleship, 1968).

In Alaska the northern pintail breeds virtually throughout the state, wherever suitable habitats occur, and it is both the most abundant and most widely distributed of Alaska's surface-feeding ducks (Gabrielson and Lincoln, 1959). In Canada it likewise has a nearly cosmopolitan breeding distribution from Banks Island to Newfoundland, perhaps being absent only from the high Arctic islands of the Northwest Territories and Nunavut, and perhaps from the still little-surveyed interiors of Ontario and Quebec.

South of Canada, the northern pintail is most abundant as a breeding species in the Great Plains and western states, from Washington south to California and eastward to Iowa and Minnesota, where it is an uncommon to occasional nester. In Wisconsin it has become an increasingly regular nester, it is an uncommon to rare breeder in Michigan (Brewer, McPeek, and Adams, 1991), and there are nesting records from Ohio, Pennsylvania, New York, Vermont, and Maine.

The breeding habitat of the northern pintail obviously varies greatly throughout its enormous geographic range. In the Arctic it is found in marshy, low country where shallow freshwater lakes occur, especially those with a dense vegetational growth near shore. It also occurs in brackish estuaries and along sluggish streams that have marshy borders (Snyder, 1957). Hildén (1964) concluded that the pintail has a psychological dependence on open landscape and thrives best in wide, open terrain with shallow waters, swamps, bog lakes, and quiet rivers. Ponds surrounded by trees appear to be avoided, but Hildén noted that either herbaceous or grassy islets are used for nesting. Munro (1944) noted that the favored breeding habitat in British Columbia is open, rolling grassland with brushy thickets and aspen copses, and adjacent sloughs or ponds. Lee et al. (1964a) has stated that in Minnesota the pintail is a bird of the prairies and is rarely found in wooded country. Keith (1961) found the highest abundance per unit of shoreline of pintail pairs on a large (21-acre) lake with a maximum depth of five feet, low shoreline vegetation, and an abundance of submerged plants.

Population. North American breeding grounds surveys in 2014 indicated a total population of 3.2 million birds, 20 percent below the long-term average (USFWS, 2014). The world population of the northern pintail includes probably well over 2 million in Europe and Asia. The average annual hunter-kill estimate in the United States during the five years 2004–08 was about 442,000, but estimates have declined precipitously from an annual high of nearly 2 million in the 1970s. Estimated total annual Canadian kills from 1990 to 1998 ranged from about 33,000 to 72,000. The estimated total US pintail kill was 471,000 in 2014, and in Canada was 22,000.

Wintering distribution and habitat. To an extent only slightly less than that of the mallard, the northern pintail is able to winter almost anywhere that a combination of open water and available food may be found. In Canada it winters north to Queen Charlotte Island on the Pacific coast and to the Maritime Provinces on the Atlantic coast (Godfrey, 1985). South of Canada it winters in varying numbers in most states but is particularly abundant in the Central Valley of California, where as many as 75 percent of the pintails in the Pacific Flyway have traditionally wintered. In recent (2000–10) Midwinter Surveys about 2.5 million pintails were estimated to be present in the United States, with 52 percent in the Pacific Flyway, 24 percent in the Central Flyway, 22 percent in the Mississippi Flyway, and 2 percent in the Atlantic Flyway. Besides the Central Valley of California, other important wintering grounds are the coastal marshes of Texas and Louisiana and the rice-growing areas of these and other southern states.

In Mexico the pintail is the most abundant species of wintering waterfowl, with the largest concentrations on the west coast and progressively smaller numbers in the interior and Gulf coast regions. Some pintails, however, continue on into Central America, and a few even reach Panama and Colombia, South America. Leopold (1959) found that the largest concentration of wintering pintails is in the delta of the Rio Yaqui in Sonora, where the birds are attracted to rice stubble. Midwinter surveys from 1982 to 2006 indicated an average of about 200,000 birds in Mexico (but considerably fewer in recent years), with notably large

Fig. 23. Male northern pintail.

numbers on the Rio Grande delta, in Sinaloa on the west coast, and through the interior highlands on wetlands in Jalisco, Chihuahua, and Michoacán.

Stewart (1962) described the optimum wintering habitats for pintails in the Upper Chesapeake region to be shallow, fresh, or brackish estuarine waters having adjacent agricultural areas with scattered impoundments. He noted that pintails also locally use estuarine bay marshes and estuarine river marshes of fresh or brackish water, as well as saltwater estuarine bay marshes.

General Biology

Age at maturity. There is general agreement that northern pintails breed in their first year of life. Seventeen of 25 aviculturists contacted by Ferguson (1966) indicated that this was true of captive birds, and it likewise seems to be generally true of wild pintails. Sowls (1955) found that 13 of 115 females banded as juveniles returned to nest at Delta, Manitoba, the next year.

Pair-bond pattern. Pair-bonds in northern pintails are renewed yearly, during a prolonged period of social display, which begins after the unisexual flocks typical of the fall period begin to merge in December (Smith, 1968).

Nest location. In one California study (Hunt and Naylor, 1955), plants that were frequently used for nesting cover were rye grass (*Elymus*), saltbush (*Atriplex*), and Baltic rush (*Juncus balticus*), although all cover types received some usage. Two other California studies (Miller and Collins, 1954; Rienecker and Anderson, 1960) indicated a preference for nesting in upland situations in relatively low plant cover. In the former study, almost 70 percent of the pintail nests they found were in plant cover no more than 12 inches high, and 16 percent were in upland situations. More than half the nests lacked concealment on at least one side, and nearly 10 percent were almost without concealment. The average distance to water was as great or greater than in any other duck species, with almost 30 percent of the nests at least 40 yards from water. Herbaceous annual weeds such as saltbush, mustard (*Brassica*), and nettle (*Urtica*) were heavily used for nest cover.

Sowls (1955) reported that about 30 percent of the pintail nests he found were more than a hundred yards from water, and some nests were farther from water than those he found of any other duck species. Keith (1961) likewise noted a high average distance of pintail nests to water (164 feet), the frequent placing of nests in sparse cover, and a tendency for pintails to use the past year's dead growth for cover. This last point is largely a reflection of the early date of nest initiation in pintails, which are among the earliest of waterfowl breeders. Pintails also frequently make their nests in shallow depressions, rendering them vulnerable to flooding by heavy rains (Sowls, 1955).

Hildén (1964) and Vermeer (1968, 1970) have investigated the tendency of northern pintails and other ducks to nest in the vicinity of gulls and terns, which might provide some protection, or at least an early warning system but risks the possibility of egg and duckling losses to gulls. Many predatory birds such as hawks and eagles have an innate avoidance of preying on birds and eggs located near their own nests, which is why some many seemingly vulnerable birds tend to nest quite close to the nests of such potentially deadly predators.

Clutch size. Pintails exhibit the same kind of variations in reported average clutch sizes as occur for mallards and, as with the mallards, this is probably a reflection of their early nest initiation and opportunities for renesting. The largest reported average clutch sizes are 9.0 for 45 "early" nests (Sowls, 1955) and 9.2 eggs reported by Miller and Collins (1954) for 41 successful nests. Average clutch sizes of slightly more than 8 eggs have been reported by Anderson (1965) and Hildén (1964).

Clutch sizes of 7 or fewer eggs have been reported by several authors. Sowls (1955) and Keith (1961) found such clutch sizes typical of late-nesting birds and considered them renests. Sowls found that about 30 percent (19 of 62 marked females) attempted to renest following initial nest losses.

Incubation period. Hochbaum (1944) reported a 21- to 22-day incubation period for incubator-hatched northern pintail eggs. Sowls (1955) reported 21 days. The shorter incubation and fledging period of pintails as compared with mallards may in large measure account for their ability to nest in more northerly latitudes.

Northern pintail, adult pair

Fledging period. Oring (1968) reported that five male northern pintails required an average of 45.8 days to attain flight, while five females averaged 40.8 days. A somewhat shorter fledging period (5–6 weeks) has also been estimated for pintails in the northern part of their breeding range (Maher and Nettleship, 1968).

Nest and egg losses. Estimates of nesting success vary greatly, with some studies indicating a success in excess of 90 percent (Miller and Collins, 1954) and others as low as about 40 percent (Hunt and Naylor, 1955). Sowls (1955) found that the northern pintail was the most persistent renester among the five species of surface-feeding ducks that he studied. He estimated that perhaps as many as 44 percent of the total pintail nests he found were the result of renesting efforts.

Miller and Collins (1954) estimated that the average hatch per successful clutch was 8.5 young, while Rienecker and Anderson (1960) found an average hatch of 7.2 eggs per successful nest. The average brood size for 70 recently hatched broods counted by the latter authors was 5.2 young. This number is nearly identical (5.3) to the average of 79 broods of comparable age reported by Ellig (1955). Skunks were a major predator of nests of pintails in his study, as well as of other ground-nesting duck species. The generally poor concealment of pintail nests probably makes them unusually vulnerable to predators that locate nests visually, such as crows, ravens, jaegers, and gulls.

With this high rate of nest losses, pintails are sometimes persistent renesters. Sowls (1955) found that 30 percent of 62 pintail renested at Delta Marsh, Manitoba, and one each a second and third time. In another study (Guyn and Clark, 2000) 55 percent of 20 Alberta females renested, and one renested twice, and in a study on the Yukon-Kuskokwim Delta 56 percent of 39 females renested. This degree of renesting in an Arctic environment having a limited breeding season is surprising.

Juvenile mortality. Because of the tendency for brood merging, counts of broods near the time of fledging fail to provide an indication of prefledging losses. Thus Rienecker and Anderson (1960) noted an average brood size of 5.2 for week-old northern pintail broods and 7.3 young per brood among broods estimated to be 5–6 weeks old. They estimated, however, that an average of 5.0 young survived to fledging, compared with an average of 7.2 hatched young per successful clutch, suggesting a prefledging mortality of about 30 percent.

Adult mortality. Sowls (1955) estimated an annual survival rate for North American pintails of about 50 percent, based on banding recoveries reported by Munro (1944). Boyd (1962) estimated a 52 percent survival rate for northern pintails banded in Russia. A massive sample of 24,370 banding recoveries of North American birds, Rice et al. (2010) determined an average annual survival rate of 75.9 percent for adult males, 65 percent for adult females, 65.3 percent for immature males, and 56.3 percent for immature females. Similar survival rates (77.6 percent for adult males, 60.2 percent for females) were reported for more than 13,000 pintails banded on the Yukon-Kuskokwim Delta (Nicolai, Flint, and Wege, 2005).

General Ecology

Food and foraging. One thorough analysis of northern pintail foods was that of Martin et al. (1951), who noted a high incidence of plant foods taken by a sample of more than 750 birds killed during fall and winter. Seeds of bulrushes (*Scirpus*), smartweeds(*Polygonum*), the seeds and vegetative parts of pondweeds (*Potamogeton*), wigeon grass, (*Ruppia*), and a variety of other native and cultivated plants were present in these samples. Bulrushes and pondweeds are also important summer foods for flightless birds, judging from a study by Keith and Stanislawski (1960). Stewart (1962) noted that the foods of 32 pintails shot in the Chesapeake Bay region had varied with the habitats utilized. Birds taken near agricultural fields showed corn and weed seeds associated with cornfields; those shot in estuarine bay marshes had a variety of seeds of submerged, emergent, and terrestrial plants and only a limited amount of corn; and those from estuarine river marshes and estuarine bays had no corn present at all.

Munro (1944) believed that, unlike mallards, pintails will not feed in cornfields where water is not immediately available in the field, and thus field-feeding opportunities for pintails are relatively limited.

Northern pintail, alert adult male

Bossenmaier and Marshall (1958) noted that pintails in Manitoba did not field-feed as zealously as mallards, and a large percentage of them usually remained on a lake. They did, however, report that dry, cut grain fields were sometimes heavily used during fall by both mallards and pintails. Unlike mallards, pintails seem to show a greater preference for small grains than for corn and often migrate out of northerly areas when waters are still open and waste corn is still available in fields (Jahn and Hunt, 1964).

Perhaps to a greater extent than most surface-feeding ducks, northern pintails dive for their food (Kear and Johnsgard, 1968), but the depth they can reach is still unknown. Sugden (1973) reported that pintail ducklings preferred feeding in shallow water near shore, and 38 percent of the food in 144 samples was vegetable matter.

Sociality, densities, territoriality. Perhaps because of the northern pintail's tendency for breeding in dry, upland situations, its population concentration on the breeding ground never seems to be extremely high. Drewien and Springer (1969) reported that over a 16-year period pintails had an average density of 5.6

pairs per square mile in a South Dakota study area. This is close to a figure of 29 pairs seen on a 4-square-mile study area (about 7 pairs per square mile) in South Dakota reported by Duebbert (1969). When calculated according to available water area, pair density per unit area of water ranged as high as 12.6 pairs per 100 acres in Drewien and Springer's study, with these high densities occurring on temporary water areas and shallow marshes. Keith (1961) noted a five-year average of 22 pairs of pintails on 183 acres of impoundments in Alberta, or about 12 pairs per 100 acres of water.

Little evidence favoring the existence of territoriality is available for northern pintails. Munro (1944) noted that there was little hostility among male pintails sharing the same nesting area. Sowls (1955) found that pintails, as well as other surface-feeding ducks he studied, lacked definite territorial boundaries, exhibited defensive behavior in various parts of their home ranges, and commonly shared loafing sites with other pairs of their species. He noted that "defensive flights" of pintails reached a peak about the time of most early egg laying, which would represent the time that females were relatively unguarded by their mates and subject to harassment by other drakes. Sowls also noted that at least six hens nested within 200 yards of a single pond, but there was almost no evidence of aggression among these pairs.

Smith (1968) likewise observed little aggression among pintails during the breeding season and confirmed that aerial pursuit behavior is closely related to the period of egg laying. Mated males also pursued other females most strongly during the time that their own mates were laying eggs. In fact, mated males were more likely than unmated ones to chase females, since unmated males more commonly responded with courtship behavior. Smith questioned a territorial interpretation of these flights and instead suggested that they tend to disperse nesting females and perhaps also ensure the fertilization of females during the egg-laying period.

Interspecific relationships. There is no definite evidence of competition between northern pintails and other duck species for nest sites or other habitat requirements. Northern pintails do exhibit a strong tendency to nest in the presence of gulls or terns (Hildén, 1964; Vermeer, 1968, 1970). Anderson (1965) also reported on ducks nesting in the vicinity of gulls, and indicated that 31 percent of 107 nests found near gull colonies were pintail nests.

Northern pintails have the usual array of egg and duckling predators, and at times seem to suffer fairly high nest losses to them (Ellig, 1955; Anderson, 1956), probably because their nests are often poorly concealed in relatively low vegetation. Many mammals are significant nest predators, especially coyotes, red foxes, and skunks in the northern prairies of the Dakotas, as are birds such as black-billed magpies and American crows.

General activity patterns and movements. The northern pintail follows a daily activity pattern that is quite similar to that described for the mallard, and indeed the two species often migrate and forage together. Pintails are exceptionally strong fliers and sometimes undertake movements of remarkable length. Chattin (1964) noted that pintails that had been banded in Alaska or elsewhere in North America have been sometimes recovered in the drainages of the Anadyr, Kolyma, and Lena Rivers of Russia, 2,000 miles or more from continental North America. Low (in Aldrich et al., 1949) described an apparent counterclockwise migration route of pintails, which sometimes move southward out of Canada through the Dakotas, westward to California, south into Mexico, and make a return spring flight through the Central and Mississippi Flyways of interior North America.

Social and Sexual Behavior

Flocking behavior. During the fall migration flight there is a surprising separation of ages and sexes in migratory flocks arriving at wintering areas, and apparently a certain degree of sexual separation persists into early winter. Smith (1968) noted large flocks of males and smaller flocks of hens in Texas during early December, followed by mixed flocks later in the month. Pair formation evidently proceeds relatively rapidly. Smith did not indicate the rate of pair formation, but at least in Bavaria about 90 percent of the females are mated by the end of February. Early flocks arriving at the breeding grounds of southern Manitoba are of paired birds, and Sowls (1955) noted that such early arrivals contained a mixture of mallards and pintails, averaging about 12 birds per flock.

Following the breeding season, and particularly after the postnuptial molt, northern pintails again begin to gather in fairly large flocks in preparation for the flight southward. Where they raft on large lakes during the hunting season, they may resort to feeding in shallow waters or on land either at night or after legal shooting hours.

Pair-forming behavior. As noted, pair-forming behavior begins on the wintering ground and is virtually completed by the time the birds have completed their spring migration. Northern pintails seem to have a moderately disproportionate sex ratio favoring males, suggesting a higher mortality rate among females. Thus, during spring migration only a few females, but many males, remain unpaired, and intense aquatic and aerial courtship activity is a prominent feature of spring pintail flights.

Male northern pintails exhibit a diverse array of aquatic courtship displays (Smith, 1968; Johnsgard, 1965, see Fig. 24), but their actual significance in the formation of pairs remains obscure. Smith noted that during aerial courtship a female sometimes indicates her preference among males by shifting in his direction, and when on water the combination of female inciting and the preferred male turning-of-the-back-of-the-head appears to be a critical factor in the formation of individual pair-bonds (Johnsgard, 1960, 1965). Smith likewise noted that when a preferred male turned the back of his head toward the female, she often responded with inciting and following him.

Copulatory behavior. Copulation is preceded by the mutual head-pumping behavior typical of surface-feeding ducks. After treading is completed, the male normally performs a single "bridling" movement similar to that of mallards but does not follow it with the usual nod-swimming. Turning-of-the-back-of-the-head and "burping" have also been observed following copulation (Johnsgard, 1965).

Nesting and brooding behavior. Female northern pintails normally lay their eggs shortly after sunrise (Sowls, 1955). Eggs are laid at the rate of one per day, and incubation begins with the last egg. The nests

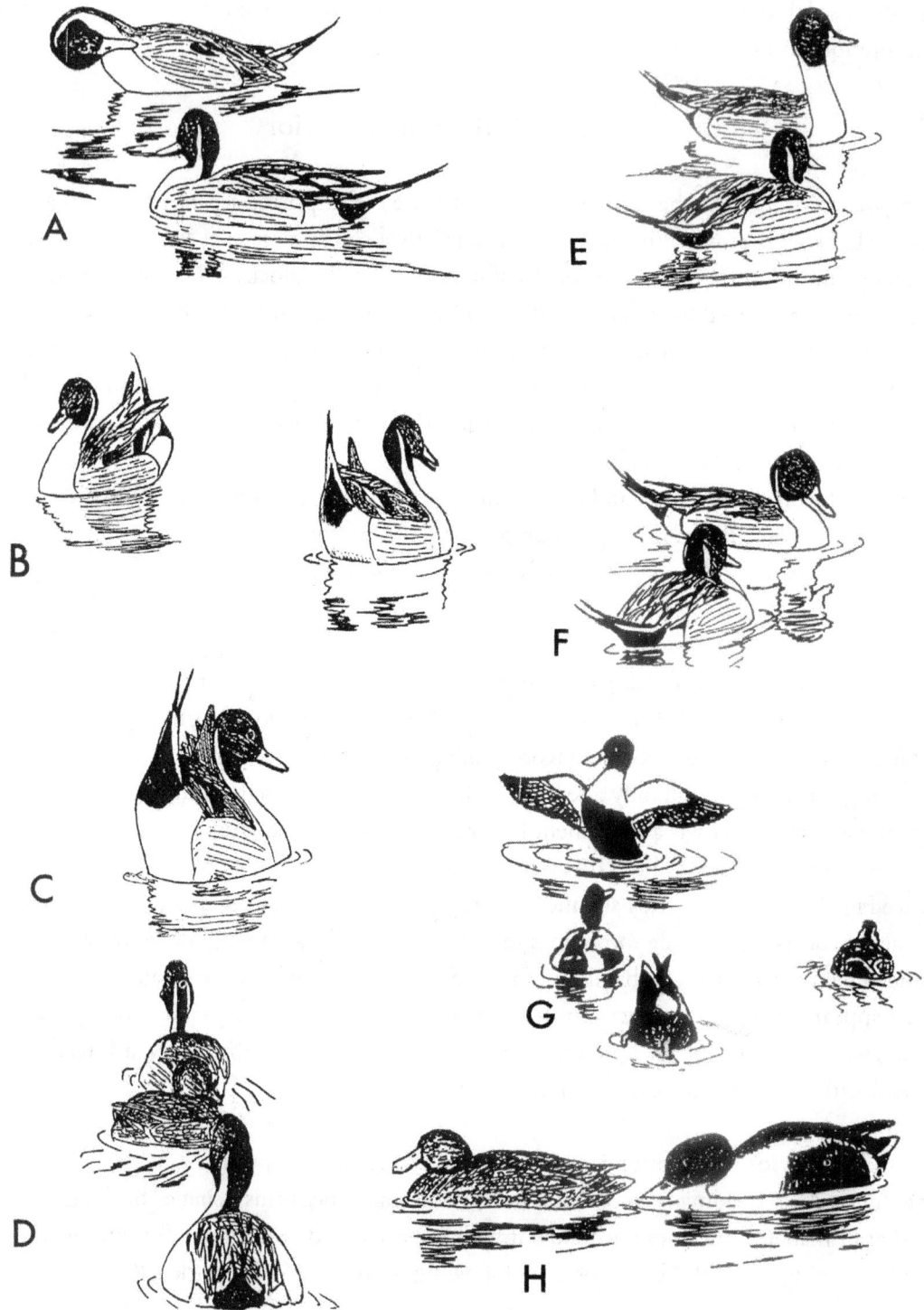

Fig. 24. Sexual behavior of northern pintail (A–F) and northern shoveler (G–H), including (A) grunt-whistle, (B–C) head-up-tail-up, (D) inciting and turning-of-the-back-of-the-head, (E–F) chin-lifting, (G) wing-flapping and tipping-up, and (H) mock-feeding.

are often so poorly concealed that the eggs may be hidden only by the usually plentiful down lining. The male may perhaps normally desert his mate only a few days after incubation begins (Sowls, 1955). An indication of the length of the pair-bond after incubation begins is provided by Smith, who noted that five of six renesting pintails remained with their original mates during renesting attempts that had resulted from initial nests being destroyed up to the twentieth day of incubation.

Following hatching, the female typically has to move her brood a considerable distance to water, and pintail broods appear to be among the most mobile of surface-feeding ducks. Sowls (1955) reported that one female pintail moved her brood 800 yards within the first 24 hours after hatching. Female pintails are among the most persistent of all surface-feeding ducks in the defense of their broods (Bent, 1923), and the seemingly low brood mortality rate of this species is perhaps a reflection of this fact.

Postbreeding behavior. By the time most females are incubating, groups of male northern pintails begin to gather in favored molting areas, such as around shallow tule-lined sloughs and marshes. Sowls (1955) determined the flightless period for male pintails to be 27 to 29 days. Males are usually flying again by early August, and females are probably able to fly by the end of that month or early September. It seems probable that tundra-breeding pintails might migrate some distance southward before undergoing their postnuptial molt, since the frost-free season would not otherwise allow the female to rear a brood before beginning her flightless period.

Garganey
Anas querquedula Linnaeus 1758

Other vernacular names. Garganey teal

Range. Breeds in Britain and from Scandinavia east across Eurasia to Sakhalin, Kamchatka, and the Commander Islands. Winters in southern Europe, northern and tropical Africa, India, and southeastern Asia, south to the southern Malayan Peninsula. Stragglers occur in North America, often along the Pacific coast.

Subspecies. None recognized.

Measurements. *Folded wing:* Delacour (1956): Males 187–198 mm; females 165–194 mm. Owen (1977): Adult males average 200.6 mm; females average 191 mm.

Culmen (bill): Delacour (1956): Males 35–40 mm; females 34–39 mm. Owen (1977): Adult males average 40.0 mm; females average 38.7 mm.

Weights (mass). Bauer and Glutz (1968): 37 males (September), average 402 g (maximum 542 g); 47 females (August), average 381 g, maximum 445 g. Owen (1977): Adult males average 359 g; adult females average 338 g.

Identification

In the hand. *Males* not in eclipse exhibit a whitish superciliary line extending down the back of the neck, elongated scapulars ornamented with glossy black and white stripes, and blackish spots or bars on the brown breast and tail-coverts. *Females* have a longer (at least 34 mm) and wider bill than the green-winged teal and show a more definite pale superciliary stripe and whitish cheek mark than either green-winged or blue-winged teal females. Both sexes have grayish upper wing-coverts, a green speculum bordered narrowly behind and more broadly in front with white, and bluish gray bill and feet.

In the field. Females cannot safely be identified in the field, and the few North American records would demand specimen identification of females. Males in nuptial plumage are so distinctive, with their rich brownish head and white head-stripe, their scaly brown breast, gray sides, ornamental scapulars, and spotted brownish hindquarters, that field identification may be possible. In flight they most resemble blue-winged teal, having similar underwing coloration but grayish rather than bluish upper wing-coverts. The voice of the male is a mechanical wooden rattling note, like that of a fishing reel. The female has an infrequent, weak, quacking voice.

Fig. 25. Male garganey.

Age and Sex Criteria

Sex determination. The somewhat brighter speculum pattern of the male, and the pale bluish gray forewing color, in contrast to the female's more brownish upper wing surface, should serve to distinguish males even when in eclipse plumage. At that time the males also reportedly have purer white throats and underparts (Delacour, 1956).

Age determination. Immatures of both sexes resemble adult females, but their underparts are more spotted and finely streaked (Kear, 2005). Notched juvenal tail feathers are probably carried for much of the first fall of life. In their absence, worn tertials from the juvenal plumage should be found to recognize first-year birds.

Occurrence in North America

During the 1970s the inclusion of the garganey on the list of North American waterfowl had rested on the fragile evidence of several sight records. These included three separate sightings in the Aleutian Islands and sightings of individuals in North Carolina, Alberta, and Manitoba. It was not until 1974 that the first North American specimen was obtained, on Buldir Island in the Aleutian Islands. By 2007 there had been four documented Texas records (Lockwood and Freeman, 2014), several records for British Columbia, and at least two each for Idaho, Oregon, and Washington. At that time garganeys had been reported from at least 30 states and seven Canadian provinces. A garganey photographed in Newfoundland in May, 2009, was the third one for that province.

Garganey, adult male

A recent (2016) eBird (NatureServe) map indicated at least ten California sight records plus four in Kansas; three in Colorado, New Mexico, Montana, and Washington; two in British Columbia, Minnesota, Oklahoma, and Texas; and one each in Alabama, Alberta, Arkansas, Indiana, Illinois, Iowa, Manitoba, Missouri, Oregon, Tennessee, Vermont, and Yukon Territory.

Blue-winged Teal
Anas discors Linnaeus 1766

Other vernacular names. Bluewing, summer teal, teal

Range. Breeds from British Columbia east to southern Ontario and Quebec, south to California and the Gulf coast, and along the Atlantic coast from New Brunswick to North Carolina. Winters from the Gulf coast south through Mexico, Central America, and South America, sometimes to southern Chile and central Argentina.

Subspecies. *A. d. discors* L.: Western Blue-winged Teal. Breeding range as above except for the Atlantic coast.

A. d. orphna Stewart and Aldrich: Atlantic Blue-winged Teal. Breeds along the Atlantic coast from southern Canada to North Carolina. Of questionable validity; not recognized by Delacour (1956).

Measurements. *Folded wing:* Males 180–196 mm, females 175–192 mm. Kear (2005): Males, average of 50, 187 mm; females, average of 31, 180 mm.

Culmen (bill): Delacour (1956): Males 38–44 mm, females 38–40 mm. Rohwer, Johnson, and Loos (2002): 33 males, average 40.1 mm; 18 females, average 39.1 mm.

Weights (mass). Nelson and Martin (1953): 105 males, average 0.9 lb. (408 g), maximum 1.3 lb. (589 g); 101 females, average 0.8 lb. (362 g), maximum 1.2 lb. (543 g). Jahn and Hunt (1964): 87 adult and immature males, average 1.0 lb. (453 g), maximum 1.3 lb. (589 g); 77 adult females, average 0.88 lb. (397 g), 216 immature females, average 0.94 lb. (425 g).

Identification

In the hand. Blue-winged teal can be easily distinguished in the hand from all other North American ducks except perhaps the cinnamon teal. Any teal-like dabbling duck with light blue upper wing-coverts, a bill that widens only slightly toward the tip, and an adult culmen length of less than 40 mm is probably a blue-winged teal.

Males in nuptial plumage exhibit a white crescent on the face and white on the sides of the rump, but no cinnamon-red body color. *Females* of blue-winged and cinnamon teal have overlapping measurements for both bill length and bill width, but the cinnamon has a slightly longer culmen (see cinnamon teal account) and has soft flaps over the side of the mandible near the tip, producing a semi-spatulate profile when viewed from the side. Additionally, female blue-winged teal almost always have an oval area at the base of the upper

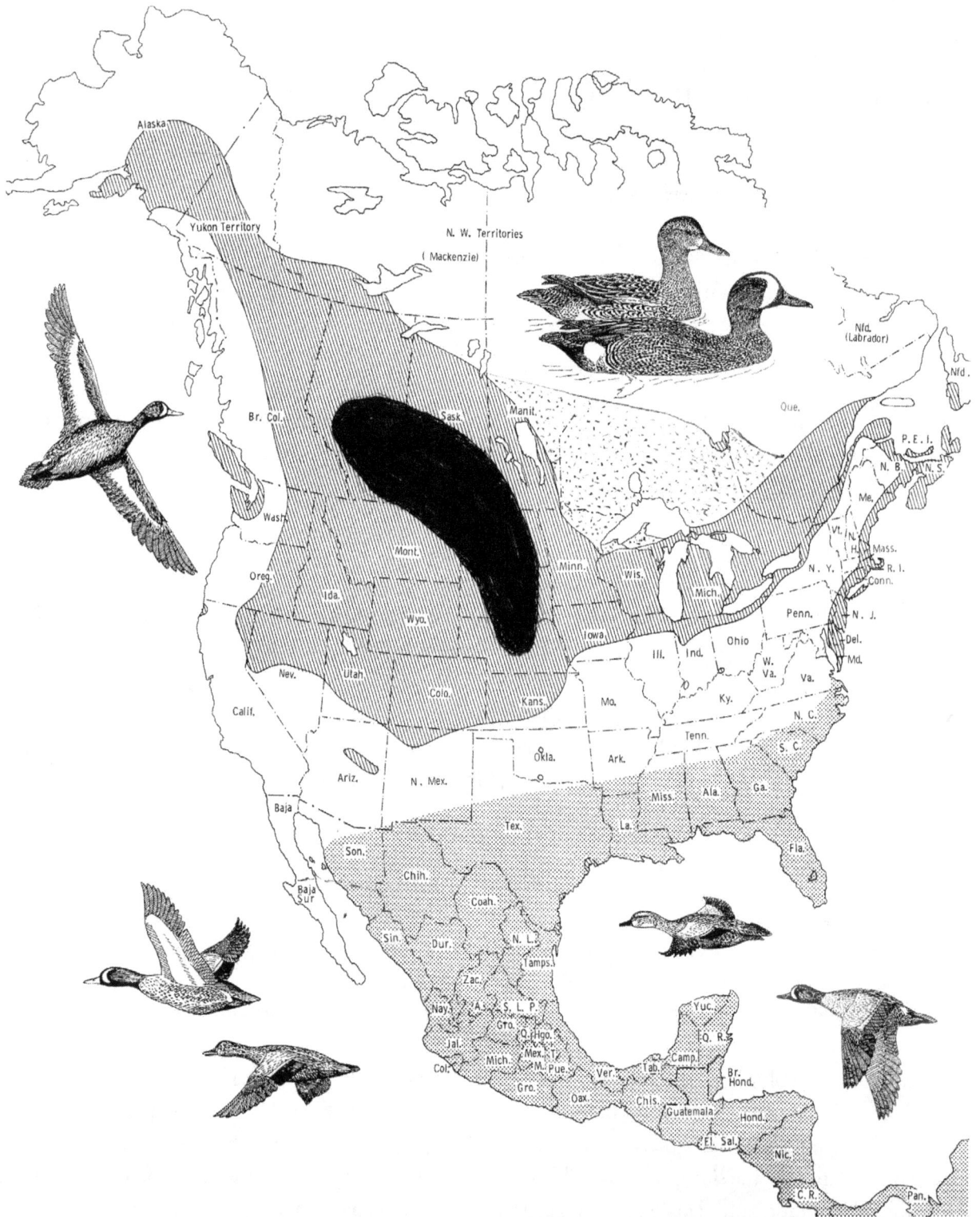

The breeding (diagonal hatching, with denser concentrations inked), wintering (shaded), and acquired or marginal (stippled) range of the blue-winged teal.

mandible that is free of tiny dark spotting and thus appears light buffy to whitish, compared with the rest of the more brownish face. The same is true of the chin and throat, although the contrast is not quite so apparent. Stark (1979) summarized morphological differences between them, and advice on distinguishing the two in life was provided by Wallace and Ogilvie (1977).

In the field. On the water, blue-winged teal appear as small dabbling ducks with dark bills and generally brownish body coloration, the white facial crescent and lateral rump spot of the male being the only conspicuous field marks. Females have rather uniformly brown heads, without strongly blackish crowns or eye-stripes but with a whitish or buffy mark just behind the bill. The bluish upper wing-coverts are normally invisible on the water, but in flight these show up well and alternately flash with the under wing-coverts, which are white except for a narrow anterior margin of brown. The call of the male is a weak, whistling *tsee* note, infrequently heard except during courtship. The female has a high-pitched quacking voice and a poorly developed decrescendo call of about three or four notes, muffled at the end.

Age and Sex Criteria

Sex determination. The presence of pale cinnamon body feathers with black spotting indicates a male except during eclipse plumage. At any time, a strongly iridescent green speculum indicates a male, whereas females have a dull green speculum. Males have white-tipped greater coverts, whereas in females these coverts are heavily spotted with dark (Carney, 1964).

Age determination. The presence of any notched tail feathers indicates an immature bird of either sex. In males, the tips of the greater secondary coverts of immatures often have dark spots, which are usually lacking in adults. The tertial coverts of immature males are narrow, pointed, and often edged with light brown, whereas in adults this is not the case. Indications of an immature female are frayed or wispy tips in the tertials, narrow greater tertial coverts that are sepia rather than greenish black, and more rounded feathers with tan edges (Carney, 1964).

Distribution and Habitat

Breeding distribution and habitat. The breeding range of the blue-winged teal is surprisingly extensive, considering its unusual sensitivity to cold weather. Although Hansen (1960) noted that the species is a regular breeder in the Tetlin area, it is rare as a breeding species in eastern and southeastern Alaska (Rowher, Johnson, and Loos, 2002).

The species breeds across most of the southern part of Canada (Godfrey, 1985) from Victoria, British Columbia, and the southern Yukon eastward to the Maritime Provinces and western Newfoundland (Tuck, 1968). Except in the Prairie Provinces, however, the blue-winged teal is not an abundant breeder anywhere in Canada. Probably the vicinity of Great Bear Lake represents the northern limit of common breeding in

Canada, and east of Manitoba breeding also becomes increasingly infrequent (Bennett, 1938). However, it does breed east to southern Newfoundland (Rowher, Johnson, and Loos, 2002).

In the United States south of Canada, the blue-winged teal breeds from the Pacific to the Atlantic coasts but has its distributional center in the marshes of the original prairies. Besides being one of the most abundant breeding species in North and South Dakota, it constitutes nearly half of the breeding duck populations of Minnesota (Lee et al., 1964a) as well as Wisconsin (Jahn and Hunt, 1964) and is the commonest of Iowa's breeding ducks (Musgrove and Musgrove, 1947). Although in the western states the blue-winged teal is outnumbered by the closely related and similar cinnamon teal, by the 1960s it had pioneered new breeding areas from British Columbia to California (Wheeler, 1965). It then also bred locally in central Arizona, New Mexico, and Oklahoma, and in Texas it was a local breeder along the Gulf coast. By the 1970s it had become locally common in Louisiana and had occasionally bred in Alabama.

On the Atlantic coast it breeds regularly from coastal Maine southward through Massachusetts, New York, New Jersey, Delaware, and Maryland. In Virginia it breeds coastally throughout; in North Carolina it breeds south at least to the Cape Hatteras region and regularly at Pea Island National Wildlife Refuge. There is occasional breeding as far south as central Florida.

Breeding surveys between 1955 and 2010 over the traditional survey routes indicated that the greatest density of blue-winged teal were then breeding in southern Saskatchewan (28 percent), followed by North Dakota (20 percent), South Dakota (19 percent), southern Alberta (13 percent) and southern Manitoba (8 percent). In the 2010 traditional survey route there were an estimated 6.4 million breeding blue-winged teal, with 61 percent in the eastern Dakotas, 21.5 percent in southern Saskatchewan, 8 percent in southern Alberta and Manitoba, and 3 percent farther north, from Alaska to western Ontario (Baldassarre, 2014).

The preferred nesting habitats of blue-winged teal are marshes in native prairie grassland, with true or tallgrass prairies of greater importance than the drier mixed prairies to the west (Bennett, 1938). Other grassland habitats used are the bunchgrass prairies of the Pacific Northwest, locally wet areas on the dry western plains, and, to a more limited extent, coastal prairies or marshes. Stewart (1962) noted that in the Chesapeake Bay area, breeding populations are mostly restricted to areas having fairly extensive salt-marsh cordgrass (*Spartina*) meadows with adjoining tidal ponds or creeks. Drewien and Springer (1969) reported that although larger ponds received heavy use by pairs prior to nesting, small and shallow marshes had the highest use by blue-winged teal during the nesting season. However, Sowls (1955) and Keith (1961) found that a variety of lake, pothole, and flooded ditch types were used by breeding birds. Glover (1956) found high nesting densities in bluegrass (*Poa*) and sedge (*Carex*) meadows with interspersed shallow sloughs having little open water.

Population. North American breeding grounds surveys in 2014 indicated a total population of 8.5 million birds, 75 percent above the long-term average of 4.9 million (USFWS, 2014). The average annual hunter-kill estimate in the United States for combined blue-winged and cinnamon teal during the five years 2004–08 was about 870,000 birds, but annual estimates have been quite variable and might reflect the influence of recent special early teal-hunting seasons. The 2014 US estimate was 1.15 million killed. Estimated total

Fig. 26. Male blue-winged teal, in flight.

annual Canadian kills from 1990 to 1998 ranged from about 22,000 to 53,000 and were 52,000 in 2015. The range map's dashed lines indicate some apparently expanded breeding regions, and the dotted line indicates recently expanded wintering regions, which have tended to move north with the global warming trend and ecological changes in coastal marshlands.

Wintering distribution and habitat. To a greater extent than any other North American duck species, the blue-winged teal migrates out of the colder portions of North America and moves to both Central and South America. Only a few hundred thousand teal were counted during winter surveys within the limits of the United States in the 1970s. Nearly 80 percent of these were in the Mississippi Flyway, primarily in coastal Louisiana, where they had been abundant since a 1957 hurricane greatly increased their food supply (Hawkins, 1964). Between 2000 and 2010 over 250,000 blue-winged and cinnamon teal were wintering north of Mexico, with 53 percent in the Mississippi Flyway, 36 percent in the Central Flyway, 6.5 percent in the Atlantic Flyway, and 2.5 percent in the Pacific Flyway. Although relatively few birds wintered there in the mid-1900s, Stewart (1962) noted that the teal's preferred wintering habitat in Chesapeake Bay consisted of brackish estuarine bay marshes. Louisiana's coastal marshes supported 99 percent of those in the Mississippi Flyway (Baldassarre, 2104). The Texas Gulf coastal wetlands supported nearly 100 percent of the teal in the Central Flyway, the vast majority of which would be blue-winged teal, as cinnamon teal are uncommon to rare winter visitors to the Gulf coast in Texas.

Several hundred thousand wintered each year in Mexico in the 1970s, where they then were the fourth most abundant wintering species of waterfowl and especially prevalent along the Gulf coast. Winter surveys of the Gulf coast of Mexico from 1978 to 2006 indicated a similar average of about 234,000 teal, three-fourths of which were found in the wetlands of Tabasco, Campeche, and Yucatan. Similar surveys from 1981 to 1994 on the Pacific coast revealed that the majority of blue-winged/cinnamon teal were wintering in coastal wetlands of Sinaloa and Nayarit (Baldassarre, 2014).

Blue-winged teal also winter on the islands of the Caribbean, especially Cuba, and throughout the Central American countries. In Puerto Rico, blue-winged teal inhabit freshwater lagoons with cattail and sedge cover, and small, open pools in the midst of dense mangrove swamps in salt or brackish water (Bennett, 1938). As far south as Panama the blue-winged teal is the most common of the wintering North American waterfowl. The species also has been recorded in the winter months over most of South America, with records extending at least as far south as Uruguay; the vicinity of Buenos Aires, Argentina; and Coquino Province of Chile.

General Biology

Age at maturity. Fourteen of 20 aviculturists contacted reported that captive blue-winged teal bred when a year old (Ferguson, 1966). Dane (1966) noted that at Delta, Manitoba, almost all first-year females initiated their first clutches before June 4 (or not significantly later than did older females.

Pair-bond pattern. Pair bonds are renewed each year during winter and early spring. The percent of females that may remate with males of the past year is still unknown but probably low, considering the long migratory routes, fairly high mortality rates, and a probable differential sex migration during fall involving an early departure of males (Jahn and Hunt, 1964).

Nest location. Bennett (1938) noted that in a sample of more than 300 nests, bluegrass (*Poa*), slough grass (*Spartina*), and alfalfa (*Medicago*) were of descending importance as sources of nest cover and that pure stands of bluegrass received the highest nesting use. Burgess et al. (1965) found that bluegrass cover accounted for 40 percent of 111 nests, with alfalfa and mixed native grasses being second and third in importance, respectively. Glover (1956) also reported a high usage of bluegrass or sedge meadows for nesting cover in Iowa. In Minnesota, alfalfa is used less for nesting than in Iowa, apparently because of its delayed growth. Dry sites in undisturbed grasses or lightly grazed pastures are preferred, with the average vegetation heights about 12 inches (Lee et al., 1964a).

Teal seem to accept nesting cover that ranges from about 8 to 24 inches high at the time of nest initiation. They avoid unusually tall cover (Bennett, 1938) and steep slopes. Depending on the topography, the nests may be situated within a foot or two of the water level (Glover, 1956) or may average as much as ten

Blue-winged teal, adult pair swimming

feet above the water level (Burgess et al., 1965). However, nests are usually within a quarter mile of water, and in one study (Glover, 1956) they tended to be about halfway between water and the highest surrounding point of land.

Clutch size. The highest reported average clutch sizes are 10.97 eggs for 100 Manitoba nests initiated before June 4 (Dane, 1966), 10.6 eggs for 54 Manitoba nests completed by June 15 (Sowls, 1955), and 10.3 eggs for 126 nests in Minnesota (Lee et al., 1964a). Eggs are laid at the rate of one per day. There is a decline in clutch size among later nests, with Sowls (1955) reporting an average clutch of 8.8 eggs in late nests, Glover (1956) noting an average clutch of 6.4 eggs in 48 apparent renests, and Bennett (1938) finding an average of 4.3 eggs in 27 renesting attempts.

Although Sowls (1955) found the incidence of renesting fairly low among blue-winged teal in Manitoba, a more recent study by Strohmeyer (1968) indicated that 35 percent to 40 percent of the unsuccessful females attempted to renest, and in certain years or situations the renesting rate may exceed 50 percent.

None of the individually marked first-year females renested, but 50 percent of the older ones did so. The hatching success and brood survival rate were similar among initial nests and renests, although the clutch sizes of renests were appreciably smaller than the original clutches, especially those which were not begun immediately after the loss of the first nest.

Incubation period. Glover (1955) and Bennett (1938) reported the incubation period to be 21 to 23 days, based on their observations in Iowa. Dane found a slightly longer average incubation period of 23 to 27 days for wild females in Manitoba. Among15 clutches that were incubated artificially the average period was 24.3 days.

Fledging period. Hochbaum (1944) reported a fledging period of 38 to 49 days, or about the same as the six-week period reported by Bennett (1938). Weller (1964) reported a 39- to 40-day fledging period.

Nest and egg losses. Bennett (1938) noted a 60 percent hatching success for 223 Iowa nests, compared with a 21 percent success for 173 nests studied in the same area by Glover (1956). Lee et al. (1964b) reported a 35 percent hatching success for 257 nests in Minnesota. He noted that the average size of 28 hatched clutches was 9.4 eggs, and the average size of newly hatched broods was 7.6 young. Jahn and Hunt (1964), summarizing nine studies, found that an estimated average of 49 percent of the females succeeded in producing broods. A large number of predators or scavengers are responsible for nest and egg destruction, including crows, skunks, ground squirrels, badgers, mink, and probably others (Bennett, 1938). Egg destruction by weasels was reported by Teer (1964). Mowing and flooding also contributed to nest losses, and mowing in hayfields is sometimes a serious source of nest losses.

Juvenile mortality. Brood counts of older broods are poor estimates of prefledging brood mortality, because of brood mergers and the occasional loss of an entire brood. Bennett (1938), counting adult female-to-young ratios, concluded that an average of about 5.1 young (of an initial successful hatch of 9.24 young) survived to reach the migratory stage by late August. These figures are close to those of Glover (1956), who estimated that 9.3 young hatched per successful nest and that broods about 8 to 10 weeks old averaged 5.16 young per female. A prefledging mortality of about 45 percent would thus seem to represent a reasonable estimate of brood losses, assuming no brood mergers. Postfledging mortality of immatures is probably high, but few estimates are available. Geis et al. (cited by Jahn and Hunt, 1964) estimated a 77 percent annual mortality rate for immature birds. Lee et al. (1964b) estimated a 62 percent mortality for mixed-age birds during the first year after banding.

Adult mortality. Boyd (1962) calculated a 55 percent survival rate for adults. Johnson et al. (1992) estimated survival rates of 59 percent for adult males, 52 percent for adult females, and 32 percent for immature females.

Blue-winged teal, adult pair feeding

General Ecology

Food and foraging. The adult food intake of blue-winged teal is approximately three-fourths vegetable material, with a somewhat higher rate of animal materials taken during spring. Seeds are especially prominent among the plant materials, although the vegetative parts of such plants as duckweeds (*Lemnaceae*), naiads (*Najas*), pondweeds (*Potamogeton*), wigeon grass (*Ruppia*), and similar aquatic plants are also consumed (Martin et al., 1951). Bennett (1938) found that the sedge, naiad, and grass families contributed over half of the total food intake of 385 teal samples on a volume basis, whereas insects, mollusks, and crustaceans composed about 25 percent. The apparently high use of seeds by blue-winged teal, as well as by many other waterfowl, may in part be a reflection of sampling bias, resulting from the slower rate of digestion of hard seeds as compared with soft foods when both are ingested simultaneously (Swanson and Bartonek, 1970).

Blue-winged teal feed almost entirely from the surface or by tipping-up; only one observation of them diving for food seems to have been published (Kear and Johnsgard, 1968). Their small body size and restriction to foraging at or near the surface probably accounts for their strong tendency to inhabit shallow and small water areas.

Sociality, densities, territoriality. The social bonds of blue-winged teal persist through spring migration, even though the majority of the birds are paired at that time (Glover, 1956). After their arrival at the breeding grounds, the males become increasingly intolerant of one another and direct their attacks primarily toward the females of other pairs (McKinney, 1970). McKinney interpreted this as territorial defense, although most other workers have not detected the presence of true territoriality in this species. Glover (1956) obtained no data during his study to support the idea of territorial defense. Bennett (1938) described "nesting territories" and "male waiting territories" but observed no defense by males of the latter, nor did he see any females defending their nesting areas. Drewien and Springer (1969) noted that during the start of nesting activities, pairs of blue-winged teal showed intolerance for other breeding birds of their species and thus tended to disperse over the available habitat. There seems, however, to be no evidence that blue-winged teal exhibit defensive behavior relative to any area per se, but rather only defense of the female.

Nesting densities of blue-winged teal in favorable habitats seem to be among the highest of all dabbling ducks. Keith (1961) found a four-year average of 31 pairs on 183 acres of impoundments in Alberta, or an average density of 18 pairs per 100 acres. Drewien and Springer reported pair densities of 17.4 to 63.6 pairs per 100 acres on various pond types during two years of study in South Dakota. Jahn and Hunt (1964) reported six-year average densities of 4 to 22 pairs per 100 wetland acres in four geographic areas in Wisconsin. Bennett found nest density estimates ranging from as low as 1 nest per 100 acres to as high as 10 nests per acre. Glover (1956), working in the same area, reported an average nest density of 1 nest per 12.5 acres of total cover, with a maximum of 1 per 1.3 acres on a 30-acre island.

Interspecific relationships. Among the other surface-feeding ducks, only the cinnamon teal is sufficiently closely related and similar in its habitat requirements as to be a possible serious competitor for mates, food, or nesting sites. Mixed courting groups involving these two species may sometimes be seen among wild birds, and several wild hybrids have been reported, although the incidence is surprisingly low considering the similarity of the females of these species. In captivity, at least, I have seen males of each species regularly performing courtship displays to females of the other species, so evidently the primary responsibility for proper species recognition resides with the female. During interactions with cinnamon teal, the blue-winged teal were more likely to initiate hostile behavior and were more overtly aggressive (Connelly and Ball, 1984).

Predators causing nest losses in blue-winged teal are numerous and include crows, skunks, ground squirrels, minks, badgers, foxes, weasels, and no doubt others (Bennett, 1938; Glover, 1956). Some of these same predators might take ducklings, as might snapping turtles, large predatory fish, and probably some avian predators.

General activity patterns and movements. Rowher, Johnson, and Loos (2002) tabulated diurnal activity budgets for breeding, postbreeding, and winter seasons. In all three seasons foraging behavior represented the single greatest proportion of available time.

Social and Sexual Behavior

Flocking behavior. Except immediately prior to and during the nesting season, blue-winged teal are distinctly flocking birds. Broods of several families typically join together during late summer, and flocks usually consist of several hundred birds during the start of the migration period (Bennett, 1938). There is apparently an early fall departure of adult males prior to that of females and immatures (Jahn and Hunt, 1964). With the start of the hunting season, the typical flocks of 100 to 500 birds break up and reconstitute themselves into groups usually containing fewer than 30 birds. During the spring migration the flocks usually number fewer than 30 birds and often consist of only a pair or two (Bennett, 1938). Glover (1956) noted that about 60 percent of the early spring migrants reaching northern Iowa were already paired.

Pair-forming behavior. McKinney (1970) noted that most blue-winged teal wintering in Louisiana are firmly mated by mid-March. The male displays occurring during pair formation are numerous (Johnsgard, 1965; McKinney, 1970). Aerial displays are few and apparently limited to short "jump-flights" by the male toward the female, apparently to attract the female's attention. Aquatic displays consist mostly of ritualized forms of foraging ("mock-feeding," tipping-up, or "head-up and up-end") and comfort movements (shaking, preening, bathing, wing-flapping). The primary display of the female is inciting, and the male's response to it is frequently turning-of-the-back-of-the-head. As McKinney noted, this is one of the most frequent of male displays and, I believe, perhaps the most important single display in the establishment of pair-bonds.

A number of observers (e.g., Bent, 1925; Bennett, 1938) reported that much of the courtship of blue-winged teal occurred in the air. Glover (1956) made the interesting observation that a male led most of the early flights he observed, while females typically led the later ones. It is highly probable that the earlier ones he observed were indeed flights associated with pair formation, while the later ones were aerial chases of the attempted rape or "expulsion flight" type, in which males that were already paired were chasing females from the pair's vicinity or were attempting to rape them.

Copulatory behavior. As in other surface-feeding ducks, copulation is preceded by a mutual head-pumping behavior that has often been confused by earlier observers with the hostile chin-lifting or pumping movements occurring during aggressive encounters. During copulation the male firmly grasps the female's nape, and McKinney (1970) once recorded a male uttering calls softly during treading. Typically the male utters a single loud whistled *peew* or nasal *paaay* note immediately after releasing the female and assumes a rather stiff and erect body posture, with his bill pointing sharply downward (Johnsgard, 1965; McKinney, 1970).

Nesting and brooding behavior. During the egg-laying phase, females visit the nest on a daily basis to lay their eggs, usually shortly after sunrise. Egg laying may begin a few days to more than a week after the beginning of nest construction (Glover, 1956). The nest is lined with available materials, usually a mixture of bluegrass and down. In about 80 percent of 134 nests studied by Glover down was not added until at least four eggs were present.

Incubation begins within 24 hours of the laying of the last egg, and usually the nest is left once or twice a day for resting and foraging. The pair-bond of the male typically begins to wane after about three days of incubation, and he starts to associate with other such males in groups of from 3 to 35 individuals (Bennett, 1938). Females probably do not leave the nest during the last 48 hours of incubation, or at least after the process of pipping begins. Within 24 hours of hatching, the female typically leads her brood from the nest and takes them into fairly heavy brooding cover. A favorite cover is a mixture of bulrushes in water 1 to 2 feet deep. Cover containing bur reed (*Sparganium*), reeds (*Phragmites*), or cattail (*Typha*) is used much less, apparently because the plant density is too great and the tall, rank plant growth crowds out important food plants and shuts out sunshine (Bennett, 1938).

Postbreeding behavior. After deserting his mate, the male moves into suitable molting cover and soon begins his postnuptial molt. Hochbaum (1944) noted that some birds may renew their wing feathers less than 3 weeks after dropping them, but he believed that a 3- to 4-week flightless period was more typical. Shortly after regaining their flight, adult males begin to leave the breeding grounds, to be followed later by females and young.

Cinnamon Teal
Anas cyanoptera Vieillot 1816

Other vernacular names. None in general use.

Range. In North America, breeds from British Columbia and Alberta southward through the western states as far east as Montana, Wyoming, western Nebraska, western Texas, and into northern and western Mexico as well as residentially in northern and southern South America. The North American population winters in the southwestern states southward through Mexico, Central America, and northwestern South America.

North American subspecies. *Anas c. septentrionalium* Snyder and Lumsden: Northern Cinnamon Teal. Breeds in North America as indicated above.

Measurements. *Folded wing:* Delacour (1956): Males 176–194 mm; females 167–185 mm. Gammonley (1996): 44 males 180–202 mm, average 191 mm; 69 females 170–192 mm, average 182 mm.

Culmen (bill): Delacour (1956): Males 39–47 mm; females 39–45 mm. Gammonley (1996): 44 males 41–49 mm, average 44.2 mm; 69 females 40–47 mm, average 42.9 mm.

Weights (mass). Nelson and Martin (1953): 26 males, average 0.9 lb. (408 g), maximum 1.2 lb. (543 g); 19 females, average 0.8 lb. (362 g), maximum 1.1 lb. (498 g). Gammonley (1996): 44 males 315–459 g, average 383 g; 69 females 265–470 g, average 372 g.

Identification

In the hand. The rich cinnamon-red color, the reddish eyes, and the lack of white on the body distinguish the breeding male cinnamon teal from the only other teal-like duck with blue upper wing-coverts, the blue-winged teal. However, males can be recognized, even when in eclipse, by their reddish to yellowish rather than brown eyes. *Females* are much more difficult to identify. Unlike female blue-winged teal, female cinnamon teal have yellowish rather than whitish cheeks with fine dark spotting extending to or nearly to the base of the bill, eliminating the pale mark or at least making it smaller than the size of the eyes. Likewise, fine dark spotting on the cinnamon teal extends farther down the chin and throat, restricting the size of the clear throat patch.

If the bill is relatively long, with culmen length of at least 40 mm (females) or 43 mm (males); is somewhat wider toward the tip; and the soft lateral margins of the upper mandible distinctly droop over the lower mandible toward the tip, the bird is most probably a cinnamon teal. Duvall (cited by Spencer, 1953)

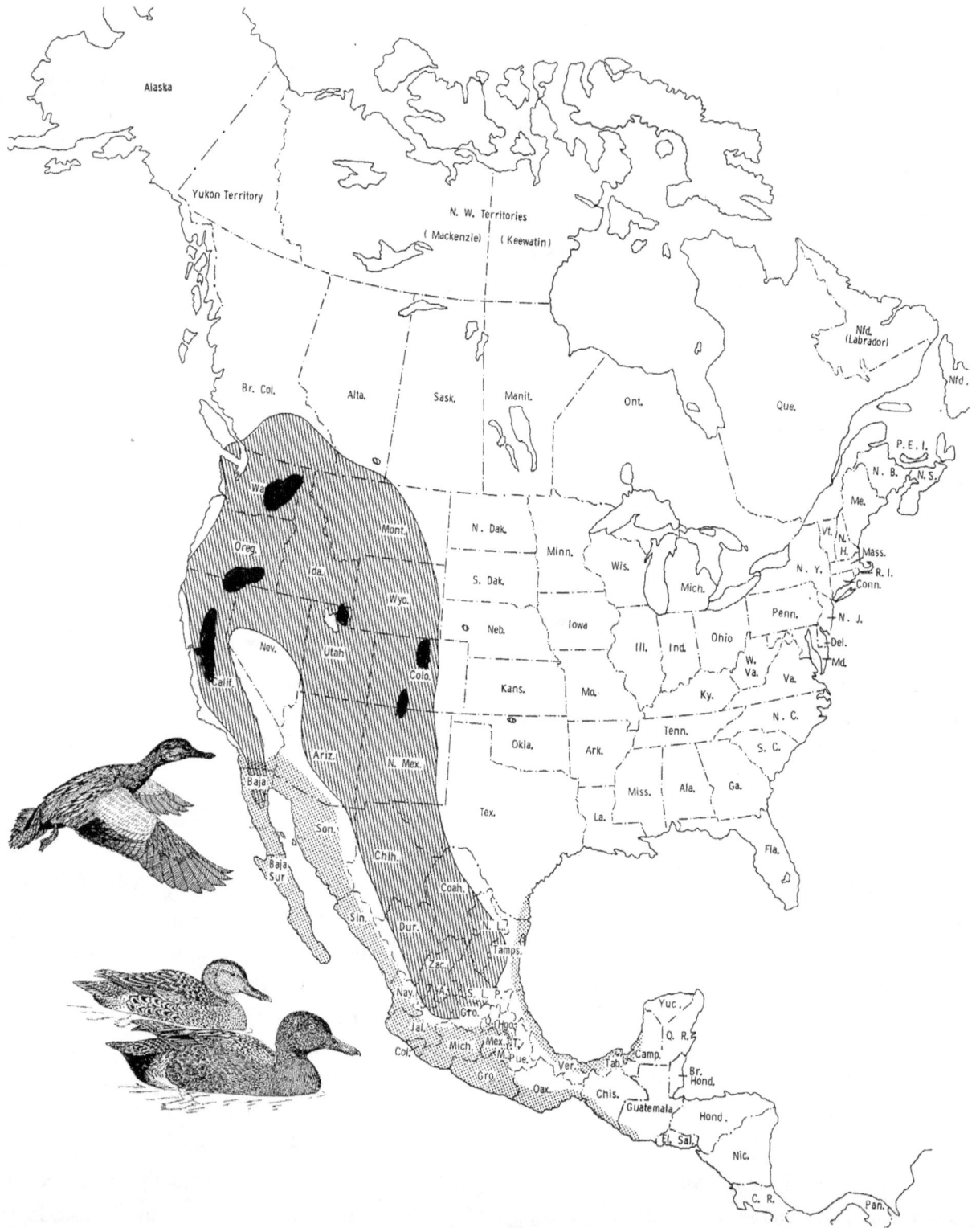

The breeding (diagonal hatching, with denser concentrations inked) and wintering (shaded) range of the cinnamon teal.

found that 26 female blue-winged teal had a maximum exposed culmen length of 41 mm, while 17 female cinnamon teal had a minimum exposed culmen length of 41 mm, with respective means of 38.7 mm and 43 mm, an approximate 10 percent difference. Rohwer, Johnson, and Loos (2002) reported the average of male blue-winged teal culmens to be 40.1 mm and females as 39.1 mm, versus cinnamon teal averages of 44.2 mm and 42.9 mm.

Wilson, Eaton, and McCracken (2012) reported that bill length measurement are 7 to 10 percent greater in the cinnamon teal (cinnamon males 42–47.9 mm, mean 46.3 mm; females 40.1–46 mm, mean 43.1 mm versus blue-winged males 37–44.7 mm, mean 41.3 mm; females 37.1–44.3 mm, mean 39.65 mm). Wing length and tail lengths are also slightly shorter in blue-winged teal, but only by using a combination of wing chord, tail length, and culmen lengths could males be identified with 96 percent accuracy by Wilson, Eaton, and McCracken, but this level of certainty was not reached with females. Significant plumage color differences exist between the two species using avian color discrimination modeling evidence, especially in breast plumage colors between the females. Stark (1979) provided additional information on morphological differences between these species.

In the field. The male's reddish underpart and upperpart coloration, and its reddish eyes allow for easy recognition. Female cinnamon teal cannot be safely distinguished from female blue-winged teal in the field except under the best conditions and by experienced observers. Their smaller cheek spot, more rusty body tone, and longer, somewhat spatulate bill are most evident when both species are side by side. The vocalizations of the females of these two species are nearly identical, but male cinnamon teal have a low, guttural, and shoveler-like rattling voice, which is uttered during courtship display. Normally, females closely associated with males of either species can be safely assumed to be of the same species, although numerous wild hybrids have been documented (e.g., Anderson and Miller, 1953; Bolen, 1978; Weseloh and Weseloh, 1979; Lokemoen and Sharp, 1981).

Age and Sex Criteria

Sex determination. The presence of reddish eyes or dark cinnamon red feathers anywhere on the head or body indicates a male. In the eclipse plumage, males can be recognized by their brighter green speculum, their yellowish red eyes, their white-tipped greater secondary coverts, or their ornamental tertials, which are pointed and blackish with buffy stripes. Immature males may lack many of these traits but are likely to exhibit at least one of them. Males acquire a reddish iris color at about eight weeks of age (Spencer, 1953).

Age determination. Notched tail feathers indicate an immature bird of either sex. Frayed or faded tertials or their coverts, which are narrow and edged with light brown, also indicate immaturity, and immature males lack the ornamental pointed and buffy-striped tertials of adults.

Distribution and Habitat

Breeding distribution and habitat. Unlike all other North American waterfowl excepting the whistling ducks and stiff-tailed ducks, the cinnamon teal has an extralimital breeding distribution in South America. In North America the northern limit of its breeding range is in western Canada, where the cinnamon teal breeds north locally to northeastern British Columbia (Baldassarre, 2014), central Alberta, and southwestern Saskatchewan.

In Washington the cinnamon teal is common east of the Cascades and occurs casually to the west; evidently it is about equally abundant with the blue-winged teal in the eastern half of the state (Yocom, 1951). In Oregon both species breed, but the cinnamon teal extends somewhat farther west and is most common in Harney, Lake, and Klamath Counties (Gilligan et al., 1994). In California the cinnamon teal commonly nests in the Tule Lake and Lower Klamath areas (Miller and Collins, 1954), in Lassen County (Hunt and Anderson, 1966), in the Sacramento Valley (Anderson, 1957), in the Merced County grasslands (Anderson, 1956), and in the Suisun marshes (Anderson, 1960). Blue-winged teal were not reported as nesting in any of these studies, but in the Lake Earl area of Del Norte County both species evidently nest, and the blue-winged teal may be the more common (Johnson and Yocom, 1966).

The cinnamon teal breeds south to Baja California, locally in Tamaulipas and as far south in Mexico as Jalisco and the central volcanic belt (Leopold, 1959; Howell and Webb, 1995). In Arizona, New Mexico, and northwestern Texas its breeding is regular but localized. The center of its breeding abundance is perhaps in Utah, where the Bear River marshes seemingly provide optimum habitat (Williams and Marshall, 1938). It breeds east to the La Poudre valley of north-central Colorado but is greatly outnumbered there by blue-winged teal, and it is a very rare breeder in the more alkaline wetlands of western Nebraska. Farther north, it regularly breeds as far east as eastern Wyoming and eastern Montana.

An analysis of breeding habitat requirements and preferences for cinnamon teal has not yet been made, but some points are evident. Like the blue-winged teal, cinnamon teal nest preferentially in fairly low herbaceous cover less than 24 inches high, preferably in grasses but with herbaceous weeds and bulrushes also locally utilized. They seem, like the gadwall, to be particularly attracted to alkaline waters, and in this respect evidently differ from blue-winged teal. Small and shallow water areas seem to receive preference over larger and deeper bodies of water. In the Potholes region of central Washington state, blue-winged and cinnamon teal pairs utilized ponds that had a surrounding grassy zone of salt grass (*Distichlis*), brome (*Bromus*), and sedges (*Carex*). Such ponds were used for nesting, while those having both open water zones and considerable emergent vegetation (mainly *Scirpus* and *Typha*) received the highest brood use (Johnsgard, 1955).

Population. The North American population was estimated at 260,000 birds in 2000 (Wetlands International, 2002). Hunter-kill figures for this species are not available because they are combined with those of blue-winged teal (see above), and both winter and breeding ground surveys likewise fail to separate blue-winged and cinnamon teal.

Fig. 27. Male cinnamon teal, preening.

Wintering distribution and habitat. Because cinnamon teal are not distinguished from blue-winged teal during winter surveys by the United States Fish and Wildlife Service, such counts are almost useless for estimating their respective winter populations. Leopold (1959) judged that of the total teal seen during the 1952 counts in Mexico, about 75 percent or more were blue-winged teal. However, cinnamon teal were noted to be prevalent among birds counted in Sinaloa and Nayarit. Areas of winter concentration were found in coastal Sinaloa and Nayarit, the southern uplands from Jalisco to Puebla, and the coast of northern Veracruz. Probably most of the cinnamon teal of North America winter in western Mexico, since the birds are apparently rare in Guatemala and are virtually unknown elsewhere in Central America. Cinnamon teal winter sparingly along the Gulf coast of southern Texas and presumably along much of the Gulf coast of Mexico, where they probably occupy habitats similar to those of blue-winged teal.

General Biology

Age at maturity. Eleven of 19 aviculturists informed Ferguson (1965) that cinnamon teal bred in captivity in their first year of life, while seven reported second-year breeding and one third-year. Comparable data from wild birds are not available, but it may be assumed that most females initially nest when a year old.

Pair-bond pattern. Cinnamon teal renew their pair-bonds each year, probably while still in their wintering areas (McKinney, 1970). In the few sex-ratio counts that have been made for this species, either males have been a surprising minority relative to females (Spencer, 1953; Johnsgard and Buss, 1956) or have constituted a slight excess (Evendon, 1952).

Nest location. In a study involving 524 nests in Utah, Williams and Marshall (1938) reported that half of the total were found in salt grass, with hardstem bulrush (*Scirpus acutus*) providing cover for another 23 percent, and most of the rest were placed in other grasses, sedges, or broadleaf weeds. In a California study, Miller and Collins (1954) reported that of forty nests found, cinnamon teal exhibited a preference for nesting on islands, using nettle (*Urtica*) cover less than 12 inches high. The nests were usually well concealed, with 70 percent being hidden from all four sides and above; all of them were within 50 yards of water, and 40 percent were within 3 yards of water. In another California study (Hunt and Naylor, 1955) involving 147 nests, ryegrass (*Elymus*) and Baltic rush (*Juncus balticus*) were primary types of nest cover, with salt grass having the next highest use.

Spencer (1953) has emphasized that specific nest cover plants may not be as important as other factors related to nest site selection. His studies at Ogden Bay and Farmington Bay, Utah, indicated a predominant use of salt grass as cover for 396 nests, whereas at Knudsen's Marsh salt grass is present in only small quantities and did not serve as cover for any of 145 nests. On the basis of cover preference calculations (usage relative to cover availability), salt grass scored much lower than many plant species occurring in trace quantities. Vegetation providing a cover height of 12 to 15 inches and good to excellent concealment was seemingly preferred, especially when such cover was close to stands of tall vegetation, such as cattails, bulrushes, or various forbs.

Clutch size. Clutch sizes for initial nests of cinnamon teal average about 9 to 10 eggs; Hunt and Naylor (1955) reported that the average size of 76 clutches from successful nests was 9.3 eggs. In a renesting study,

Fig. 28. Male cinnamon teal, in flight.

Hunt and Anderson (1965) noted that six initial nestings averaged 10.0 eggs, six second nestings averaged 8.3 eggs, and a single third nesting attempt had 9 eggs. Spencer (1953) reported an average clutch of 8.9 eggs in 104 successfully hatched nests, with very early and very late clutches tending to be smaller than those of mid-season.

Incubation period. The incubation period for cinnamon teal is reported as 24 to 25 days by Delacour (1956). Spencer (1953) observed a range of 21 to 25 days in wild cinnamon teal nests, which was supported by Gammonley (1996).

Fledging period. Spencer (1953) reported that captive-reared birds were fully feathered and probably capable of flight when seven weeks old, and Gammonley (1996) estimated a fledging period of about 49 days.

Nest and egg losses. One of the highest reported nest successes was that of Williams and Marshall (1938), who found that 84 percent of 2,655 eggs in 524 nests hatched. Hunt and Naylor (1955) found an even higher hatching success, 93 percent of 125 eggs in 1951, and 85.5 percent of 583 eggs in 1953.

Girard (1941) reported a 72 percent hatching success for 22 nests in Montana. However, Anderson (1956) found that only 20 percent of 70 nests studied in Merced County, California, hatched in 1953 and only

1.9 percent of 56 nests hatched in 1954. Most of these losses were attributed to various mammals, including dogs, cats, raccoons, skunks, and opossums. Spencer (1953) noted that skunks and California gulls destroyed 41.5 percent of 1,870 teal eggs during two years of study at Ogden Bay, Utah, where annual nesting and hatching successes were 45 and 43 percent, respectively. Brood parasitism by redheads was fairly frequent and resulted in a slight decrease in hatching success through increased nest desertion rates and in a slight decrease in average sizes of teal clutches.

Juvenile mortality. Reinecker and Anderson (1960) estimated that an average of 9.2 ducklings hatched from successful nests and that prefledging mortality reduced this number to an average terminal brood size of 6.2 young. Spencer (1953) reported average brood size reductions from about 9 ducklings shortly after hatching to 4.5 to 4.7 young for broods about ready to fledge, or approximately a 50 percent prefledging mortality, based on two years of data. No estimates of postfledging mortality rates of immature birds are available.

Adult mortality. Estimates of adult mortality rates are still limited, as relatively few cinnamon teal have been banded. Kozlik (1972) estimated an adult annual survival rate of 46 percent for birds banded in California, and 28 percent for immatures, based on small sample sizes.

General Ecology

Food and foraging. Few food analysis studies have been performed on cinnamon teal, although it seems probable that dietary differences from the blue-winged teal would be very few. Martin, Nelson, and Zim (1951) noted the seeds of bulrushes, salt grass, and sedges, and the seeds and vegetative parts of pondweeds (*Potamogeton*) and horned pondweeds (*Zannicheliia*) in summer and fall food samples. The small amount of animal materials present included mollusks, beetles, bugs, fly larvae, and the naiads of dragonflies and damselflies.

Sociality, densities, territoriality. Williams and Marshall (1938) estimated the cinnamon teal's breeding density on 3,000 acres of potential nesting cover to average 0.17 nests per acre, or nearly 110 nests per square mile. Hunt and Naylor (1955) estimated that 266 pairs of cinnamon teal were present in Honey Lake Valley in California and mostly nested in the 2,000-acre Fleming Unit of that management area, representing an approximate density of 90 pairs per square mile. Spencer (1953) calculated a nesting density of 0.18 nests per acre for 357 acres on a Utah study area, or about 120 nests per square mile.

All of these studies suggest that breeding densities of 100 or more pairs per square mile of habitat are possible among cinnamon teal, which is considerably greater than most figures available for blue-winged teal. Quite possibly the effects of crowding produced by the relatively fewer areas of marsh habitat available in the arid western states account for this apparently higher nesting density. McKinney (1970) noted that paired cinnamon teal, like blue-winged teal, restrict their activities to relatively small areas, although the home ranges of neighboring pairs tend to overlap and territorial boundaries are difficult to define. Spencer (1953) reported that most territories he observed were less than 30 square yards in area, with the nest site inside these limits or no more than 100 yards from it.

Cinnamon teal (left) and blue-winged teal (right), adult males

Interspecific relationships. The extent to which cinnamon teal and blue-winged teal might compete for food or other aspects of their habitat in areas of joint breeding is still unknown. In central and eastern Washington both species are about equally common and appear to occupy virtually identical habitats (Yocom, 1951; Johnsgard, 1955). Connelly and Ball (1984) provided comparison of their breeding ecologies and behaviors in Washington.

As with other surface-feeding ducks, a variety of mammalian and avian predators probably take eggs and ducklings, but in no case has this been proven a serious limiting factor controlling teal populations.

General activity patterns and movements. Nothing specific on activity patterns and movements is available. Spencer (1953) noted that this species is diurnal and that migrating flocks were often seen during the daytime but not at night. He also noted that social display could occur at any time during the day but was most intense before 10:00 a.m. and after 4:00 p.m. Cool and cloudy weather increased the frequency of midday display activities.

Social and Sexual Behavior

Flocking behavior. Most observers report that cinnamon teal generally are to be found in small flocks, usually consisting of paired birds (Phillips, 1924). However, this would not apply to fall flocks, since pairing has not occurred by that time. Spencer (1953) reported that the spring migrant flocks he observed were often in groups of 10 to 20 birds, while during fall the early flocks of migrating males were usually in groups of fewer than 150 birds.

Pair-forming behavior. Displays associated with pair formation probably begin on the wintering grounds when the males have regained their nuptial plumage, or roughly at the end of the calendar year. Spencer (1953) observed captive birds displaying as early as late February, but by the time the wild cinnamon teal migrants arrived in Utah during March a large percentage already appeared to be paired. The displays associated with the formation of pairs are extremely similar to those of shovelers and blue-winged teal, with ritualized forms of foraging behavior being the most conspicuous and probably most frequent displays.

As in the other two species of "blue-winged ducks," short "jump-flights" are also more prevalent than is true of the other surface-feeding ducks. McKinney (1970) was probably correct in pointing out that the presence of light blue upper wing-coverts on this group of species is evidently related to their exposure during such display flights. Inciting by females takes on a strong vertical head-pumping component, which is somewhat similar to that occurring in a precopulatory situation. The male's usual response is to perform the turning-of-the-back-of-the-head display while swimming in front of her. Very probably this display plays a major role in the formation of pairs.

Copulatory behavior. Mutual head-pumping movements precede copulation, with the tip of the bill tilted slightly downward rather than upward as in hostile encounters. After treading is completed the male may utter a single soft rattling note; he assumes a lateral posture with bill pointed downward, hindquarters and wings somewhat raised, and shakes his tail while paddling his feet (McKinney, 1970).

Nesting and brooding behavior. Females usually construct a rather simple nest of dead grasses and plant stems, with fresh green material rarely being used. They are usually shallow bowl-shaped depressions that are lined with more plant materials and down as the clutch nears completion. The first few eggs may be deposited at intervals of 1 to 3 days, while the later ones are usually at the rate of one per day, with most laying done between the hours of 8:00 a.m. and 10:00 a.m. Incubation begins within 24 hours of the laying of the last egg, and during the incubation period the female may feed for a maximum of two hours a day, usually during late afternoon. As little as seven hours may elapse between the start of pipping and the evacuation of the nest.

After hatching, the female moves her brood to rearing cover that provides adequate foraging opportunities, such as small ditches or ponds, and suitable escape cover, such as surrounding emergent vegetation.

Cinnamon teal, courting group

If suitable waterways are present, the broods may move as far as a mile in three or four days but are more likely to remain in a small area (Spencer, 1953).

Postbreeding behavior. Male cinnamon teal probably desert their mates during the early stages of incubation. Spencer (1953) did not observe any sizable groups of males during the postbreeding molting period, but by early August adult males were already beginning their southward migration. Adult males were rarely encountered after mid-September, and after mid-October the majority of the total cinnamon teal population had moved southward out of northern Utah. The rate of the southward movement is apparently rather fast, even for immature birds. One immature female, banded at Ogden Bay on July 31, was shot near Mexico City on August 15, suggesting an average minimum movement of 114 miles per day.

Northern Shoveler
Anas clypeata Linnaeus 1758

Other vernacular names. Shoveller, spoonbill, spoon-billed duck

Range. Breeds throughout much of the Northern Hemisphere, including the British Isles, Europe except for northern Scandinavia, most of Asia except for the high Arctic, and in North America from western and interior Alaska southward to California and eastward to the Great Lakes and St. Lawrence valley, with some breeding along the middle Atlantic coast.

Subspecies. None recognized.

Measurements. *Folded wing:* Delacour (1956): Males 225–245; females 220–225 mm. Owen (1977): Adult males average 241.2 mm; adult females average 224.2 mm.

   Culmen (bill): Delacour (1956): Males 62–64 mm; females 60–62 mm. Owen (1977): Adult males, average 67.5 mm; adult females, average 61.0 mm.

Weights (mass). Nelson and Martin (1953): 90 males, average 1.4 lb. (634 g), maximum 2.0 lb. (906 g); 71 females, average 1.3 lb. (589 g), maximum 1.6 lb. Bellrose and Hawkins (1947) and Jahn and Hunt (1964) (merged fall data): 21 adult males, average 1.53 lb. (694 g); 65 immature males average 1.49 lb. (676 g); 15 adult females, average 1.41 lb. (639 g); 68 immature females, average 1.34 lb. (608 g). Owen (1977): Adult males average 611 g; adult females average 556 g.

Identification

In the hand. The species' strongly spatulate bill, which has soft lateral margins near the tip that hang over the sides and obscure the long lamellae, is unique to the shoveler among North American species of waterfowl. Additionally, the light blue upper wing-coverts and the orange legs and feet are distinctive for both sexes.

In the field. Whether on the water or in the air, the long, spoonlike bill of both sexes is easily apparent, being distinctly longer than the head and destroying the otherwise fairly sleek lines of the duck. Males do not acquire their striking nuptial plumage until rather late in the winter, so that during fall most shovelers are female-like in appearance, with the enlarged bill and bluish upper wing-coverts being the primary field marks, the latter normally visible only when the bird is flying. In flight, the underwing surface is entirely white, and the underparts of females or dull-plumaged males are brownish, so that from underneath the birds distinctly resemble female mallards except for the more prominent bill.

176

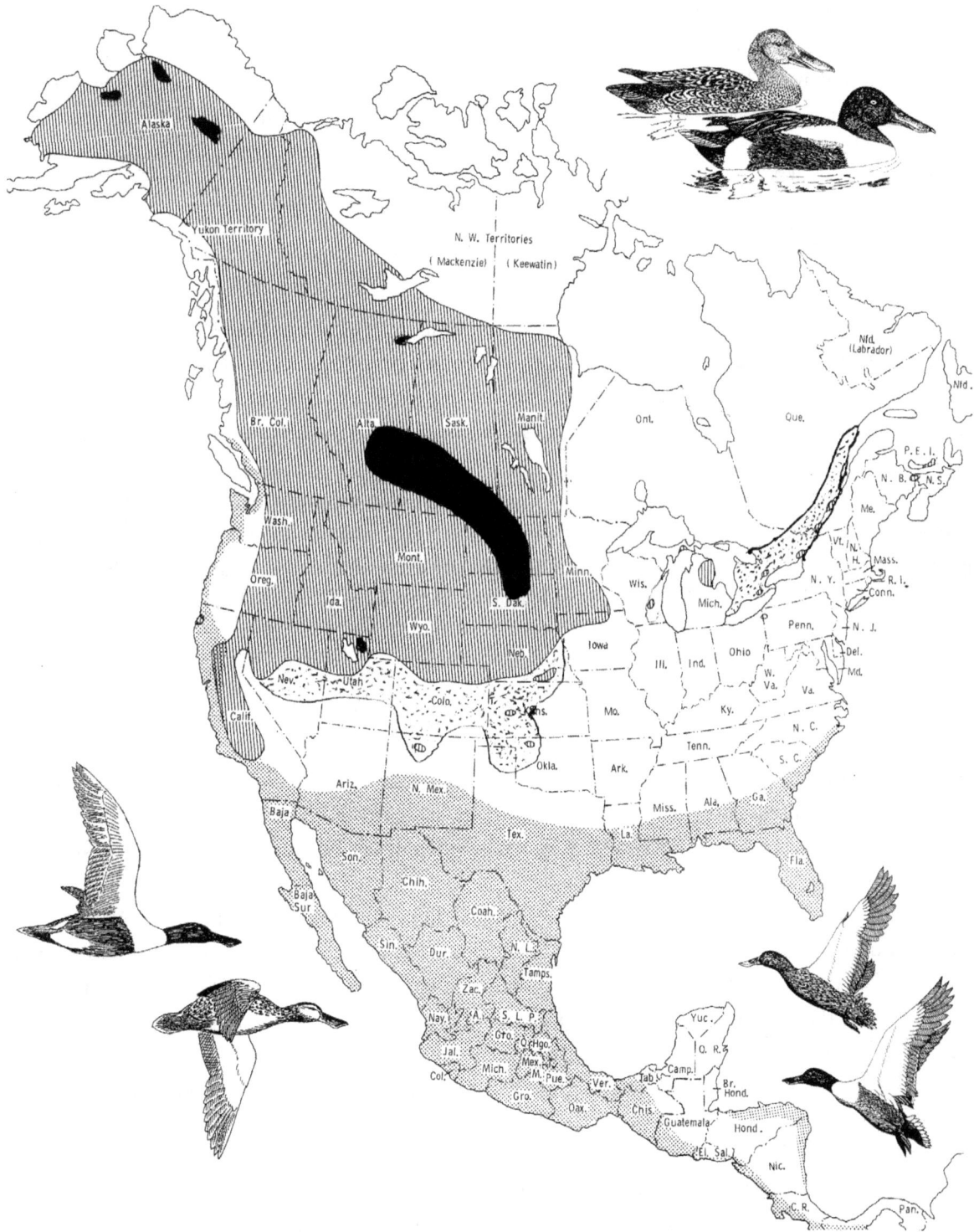

The breeding (vertical hatching, with denser concentrations inked), wintering (shaded), and acquired or marginal (stippled) range of the northern shoveler.

During late winter and spring the males acquire a white breast, a large white area between the black tail-coverts and the reddish brown sides, and an iridescent green head. At this time they are reminiscent of male mallards, except that the breast is white and the sides reddish brown, instead of vice versa. Males are quite silent except during aquatic courtship, when low-pitched rattling notes are uttered. The female has a quacking voice similar to those of cinnamon and blue-winged teal, and her decrescendo call is usually about five notes long, with the last one or two rather muffled.

Age and Sex Criteria

Sex determination. The presence of iridescent green on the head or of any pure white or chestnut brown feathers on the body indicates a male. All birds with completely noniridescent secondaries are females, but some females do show iridescence on the secondaries. Most females exhibit cream edging on the lesser and middle coverts, while males lack this or have only a few cream-edged feathers near the wrist (Carney, 1964).

Age determination. The presence of notched tail feathers indicates an immature bird, and most immatures also have small dusky spots on their greater coverts, which are lacking in adults. The presence of fading and fraying on the tertials or their coverts indicates immaturity. In immatures these are brownish or brownish black, while in adults the tertials are greenish black (males) or heavily washed with white at the tips (females), according to Carney (1964).

Distribution and Habitat

Breeding distribution and habitat. Like the other Holarctic surface-feeding ducks—the gadwall, northern pintail, northern mallard, and green-winged teal—the northern shoveler occupies a broad breeding range across most of North America. In Alaska it is generally uncommon but is most abundant on the Copper River Delta and the lakes of the Minto region.

In Canada, the shoveler as a breeding species is largely limited to the area west of Ontario and extending northward to tree line as well as westward to the coastal range of mountains in British Columbia. In Ontario and to the east it is only a very localized breeder, with most of the records from near the eastern Great Lakes and the St. Lawrence valley. It has also bred in eastern New Brunswick and on Prince Edward Island (Godfrey, 1966).

In the United States south of Canada, the shoveler breeds from central Washington southward through Oregon to south-central California, northern New Mexico, and eastward across the Great Plains to Nebraska, with localized breeding localities farther south from Kansas to northern Texas. Northern Iowa and Minnesota apparently represent the eastern limit of regular breeding by shovelers. In Wisconsin the species breeds only occasionally and in a few localities, and in Michigan the records likewise are mostly limited to a few eastern counties. Shovelers have occasionally nested in Ohio, and in New York have nested on Long Island and in the Montezuma marshes north of Lake Cayuga. They have also nested locally in New Jersey, Delaware, and at least as far south in North Carolina as Pea Island National Wildlife Refuge.

Fig. 29. Male northern shoveler.

Shallow prairie marshes represent the preferred breeding habitat of shovelers, particularly those with abundant plant and animal life floating on the surface, such as duckweeds and associated biota. Drewien and Springer (1969) reported the highest density of pairs during the nesting period on shallow marshes, with somewhat lower usage of shallow to deep marshes. Keith (1961) indicated that the highest shoveler usage in his study areas occurred on a fairly large, shallow lake with a maximum depth of five feet and with water milfoil (*Myriophyllum*) and pondweeds (*Potamogeton*) as principal submerged aquatic plants. Female shovelers leading broods seem to favor especially water areas having an abundance of pondweeds as well as waterweeds (*Anacharis*); the latter also usually supports an unusually rich associated animal life (Girard, 1939).

Hildén (1964) concluded that the nesting habitat of shovelers must include waters with open rather than wooded shores, the waters preferably being shallow, eutrophic, and with a mud bottom. Coastal shorelines that offer freshwater pools or shallow shores for feeding are acceptable, and nesting sometimes occurs on islets with gravel or polished rock shorelines. There is apparently a moderately strong attraction of shovelers to nesting gulls or terns.

Population. North American breeding grounds surveys in 2014 indicated a total population of 5.3 million birds, 114 percent above the long-term average of 2.5 million (USFWS, 2014). The average annual hunter-kill estimates in the United States during the five years 2004–08 were about 613,000 birds and apparently have been slowly increasing since the 1960s, but the estimates have been quite variable from year to year. In 2014 the estimated U.S. kill was 700,000, and the average from 1998 to 2008 was 542,000 birds. Of that total, 39 percent were from the Pacific Flyway, 38.5 percent from the Mississippi Flyway, 20 percent from the Central Flyway, and 3 percent from the Atlantic Flyway (Baldassarre, 2014). Estimated total annual Canadian kills from 1990 to 1998 ranged from about 10,000 to 27,000, and in 2014 were 22,000.

Wintering distribution and habitat. According to winter survey data of the 1970s, approximately 90 percent of the total North American wintering shoveler population of 586,000 birds then occurred in the Pacific and Mississippi Flyways, with about equal abundance in each, and the remainder in the Central and Atlantic Flyways. Since 1957 increasing numbers have wintered in Louisiana, and along the Pacific coast they commonly winter as far north as Puget Sound. In California they are abundant winter residents in the Central Valley and to a lesser extent along the coast. Between 2000 and 2010 the midwinter U.S. surveys averaged 998,000 shovelers, with 53 percent seen in the Pacific Flyway, 32 percent in the Mississippi Flyway, 14 percent in the Central Flyway, and 1 percent in the Atlantic Flyway. More than 90 percent of the Pacific Flyway shovelers were found in California, and a third of those seen in the Mississippi Flyway were in Louisiana (Baldassarre, 2014).

In Mexico the northern shoveler is outnumbered only by the pintail and lesser scaup among wintering waterfowl; it is especially abundant on the Pacific coast, where more than 200,000 birds could usually be found during the 1970s. Winter surveys in Mexico from 1981 to 2005 revealed a similar average of 213,000 shovelers along the Pacific coast. Along the Rio Grande delta and the Gulf coast of Mexico south to the Yucatan wetlands about 25,000 birds were present, and about 94,000 were wintering in the wetlands of Mexico's interior highlands (Baldassarre, 2014).

Shovelers become progressively less common on the Pacific coast through Guatemala and El Salvador and occur irregularly in Panama. It is fairly common during winter on the Caribbean slope at least as far south as Honduras, and extends northward and eastward along the Gulf coast and south Atlantic states at least as far north as Chesapeake Bay.

In the Chesapeake Bay area, transient and wintering shovelers are usually well distributed on fresh and brackish estuarine bay marshes and are generally commonest on still-water ponds subject to slight tidal variations. In saltwater situations shovelers are usually more localized and apparently prefer artificial impoundments along drainage systems (Stewart, 1962).

General Biology

Age at maturity. Eleven of 20 aviculturists reported shovelers breeding under conditions of captivity at the age of one year (Ferguson, 1965).

Pair-bond pattern. Pair-bonds are lacking in shovelers between late June and the time they again acquire their nuptial plumage, about November to December (McKinney, 1970). The incidence of remating with

Northern shoveler, adult pair

mates of the previous breeding season among wild birds is still unreported, but McKinney (1965) noted that among captive shovelers some birds re-paired with the same mate, while others chose new mates. As McKinney (1970) has emphasized, there is no evidence that polyandry is characteristic of shovelers.

Nest location. In one Montana study, Girard (1939) found that more than half of 132 nests utilized short grasses, 23 percent were hidden in tall grasses, 13 percent in thistles (*Salsola* and *Cirsium*), and the rest were under various other herbaceous or shrub covers. In a Utah study, salt grass (*Distichlis*) provided cover for 65 percent of 37 nests, with bulrushes (*Scirpus*) and various herbaceous weeds making up most of the remainder. Since the favored salt grass typically grows adjacent to water, most of the nests were located fairly near water. However, Keith (1961) found that the only other surface-feeding ducks with nest locations averaging farther from water than the shoveler were the gadwall and pintail, both of which are noted for their upland nesting tendencies. Miller and Collins (1954) verified a tendency for upland nesting by shovelers as well as a preference for nesting in grasses usually less than 12 inches high and almost never more than 24 inches high. They found that almost 30 percent of the total nests located were more than 50 yards from water, and more nests were over 100 yards from water than was true of any other species.

Clutch size. The largest average clutch size reported for shovelers is 10.7 for early nests, compared to an average of 10.1 for all 45 nests of this species that were located by Keith (1961). Hildén (1964) reported an overall average clutch size of 9.19 for 43 nests. It seems likely that the relatively low average clutch size (8.2) reported by Williams and Marshall (1938) might have reflected at least some renesting efforts, which have been documented in shovelers (Sowls, 1955) but apparently occur at a rather low incidence.

Incubation period. Clark et al. (1988) estimated the incubation period of wild shovelers to be about 28 days. This is substantially longer than the 21 to 22 days reported by Hochbaum (1944) or the 22 to 25 days estimated by Bauer and Glutz (1968).

Fledging period. Reportedly, fledging may occur as early as 39 to 40 days after hatching (Weller, 1964), but Clark et al. (1988) reported the period to be 50 days.

Nest and egg losses. Girard (1939) estimated that on two Montana wildlife refuges where predator control was practiced, 69.7 percent of 1,135 eggs successfully hatched. In a Utah study, 90 percent of 189 eggs hatched, with predation playing a minor role in nest failures. Keith (1961) estimated that 42 percent of 60 nests he found in Alberta hatched successfully, and judged that about 75 percent of the unsuccessful females attempted to renest, so that a total of 62 percent of the females eventually brought off broods.

Juvenile mortality. Girard (1939) believed that about 6 eggs per successful nest hatched in his study, and of these, 5 young typically survived to reach the "flapper" stage. McKinney (1967) noted that female shovelers often killed ducklings from other broods, but the birds in his study were unusually crowded. It is clear from brood counts such as those made by Rienecker and Anderson (1960) that under natural conditions some brood mergers do occur and terminal brood sizes may be substantially larger than brood sizes at hatching. These authors estimated that about 7 young per brood represented the actual terminal brood size in their study, compared to an observed average of 10.3 hatched young.

Postfledging mortality. Postfledging mortality rates are not yet well established. Lee et al. (1964a) estimated a 29 percent survival rate for a small sample of hatching-year birds in Minnesota, and Blums et al. (1996) estimated a 38 percent survival rate for a much larger, multiyear sample of immature females in Europe.

Adult mortality. Boyd (1962) calculated a 44 percent annual adult mortality rate for shovelers banded in Britain. Wainwright (1967) calculated a somewhat lower (37 percent) mortality rate for this species. Keith (1961) estimated an all-age mortality of 58 percent annually. Lee et al. (1964a) estimated a 59 percent survival rate for adults banded in Minnesota. Blums et al. (1996) estimated a 58 percent survival rate for a large sample of adult females banded in Europe.

General Ecology

Food and foraging. Perhaps to a greater extent than any other North American surface-feeding duck, the shoveler consumes a considerable amount of small aquatic animal life, especially forms such as ostracods, copepods, and similar crustaceans that it is able to "sieve" from the water with the long, closely spaced lamellae of its bill. Insects such as aquatic beetles, water boatmen, caddisfly larvae, naiads of damselflies and dragonflies, and small mollusks also may represent important foods at various seasons or locations. Duckweeds (*Lemnaceae*) and the vegetative parts of pondweeds (*Potamogeton*), wigeon grass (*Ruppia*), and other aquatic plants are also taken, as are the seeds of bulrushes (*Scirpus*), pondweeds, and others (Martin et al., 1951).

A limited sample of shovelers taken in the Chesapeake Bay area had consumed seeds of three-square (*Scirpus*), wigeon grass, salt grass (*Distichlis*), the vegetative parts of wigeon grass and muskgrass (*Chara*), and a variety of mollusks, crustaceans, and small fish (Stewart, 1962). During spring and summer, at least, the seeds of spike rush (*Eleocharis*) appear to be a favored food for shovelers as well as blue-winged teal and other surface-feeding ducks (Keith, 1961).

Shovelers have been observed diving for food on only a few occasions (Kear and Johnsgard, 1968), and they usually are found on waters so shallow that diving is not required. McKinney (1970) observed shovelers diving for food occasionally but noted that they predominantly feed at the surface and to a lesser extent by tipping-up.

Sociality, densities, territoriality. McKinney (1970) has stressed the high degree of hostile behavior that he observed among captive shovelers and agreed with Sowls (1955) that shovelers are the most territorial of all the North American dabbling ducks. McKinney believed that several of the shoveler's display patterns had their origins in the territorial system of shovelers. However, Hori (1963) found strong mate defense but no evidence of territoriality among wild shovelers. Poston (1969) observed little territorial behavior among wild shovelers in a fairly dense population, and it seems possible that the apparently strong territoriality noted by McKinney was an artifact of maintaining a large number of pairs (4–7) in pens of less than one acre in area. Poston (1969) found that ponds under 1.25 acres were used by only a single pair of wild shovelers, while five ponds ranging from 1.25 to 2.0 acres were each occupied by two breeding pairs. He also found that the home ranges of six pairs averaged 49.7 acres and ranged from 15 to 90 acres. On a study area of three square miles, he reported breeding densities of 11.3 and 12.7 pairs per square mile during two years of study.

Stoudt (1969), reviewing breeding density figures from five prairie study areas, noted shoveler densities of 2 to 10 pairs per square mile. It is possible that not only the rather small body size of shovelers (Goodman and Fisher, 1962) but also their strong dispersal tendencies, compared with most other dabbling ducks, are reflections of the fact that shovelers probably have to "work harder" for their food and must be able to forage over a larger area than do other surface-feeders.

Interspecific relationships. Because of their highly specialized bill form, shovelers probably compete very little with other *Anas* species for food. The cinnamon teal's bill form exhibits an incipient degree of spatulate development.

In their nest site preferences and tendency to breed along open shorelines, shovelers are similar to pintails and, to a more limited extent, gadwalls. Weller (1959) reported that the shoveler has been reported to be socially parasitized by the redhead and lesser scaup, and its eggs have been found in the nests of mallards, American wigeons, cinnamon teal, and redheads.

A variety of egg predators has been reported for shovelers, including skunks and crows (Sowls, 1955; Girard, 1939). Weasels have sometimes been known to take shoveler ducklings but are probably not generally significant influences.

General activity patterns and movements. One of the few studies of general activity patterns of shovelers is that of McKinney (1967), who reported on the breeding phase of the life cycle. He noted that during the prelaying period females inspected possible nesting cover during the morning hours, especially near dawn. Likewise, egg laying was performed during the same hourly schedule. During incubation, females always spent the early morning hours on the nest and exhibited a peak in periods away from the nest during late afternoon.

Copulations were seen at nearly all times of the day. They were observed by McKinney as early as 23 days before the laying of the first egg but diminished during the egg-laying period and were rarely seen during incubation. Male chasing activities were seen throughout the prelaying through incubation period, but only infrequently did males attempt to rape strange females, and they were rarely successful. There seemed to be no correlation between time of day and frequency of chases by males.

Social and Sexual Behavior

Flocking behavior. Regrettably, little has been written on flock sizes of shovelers, which are of interest because they would shed light on the question of possible intraspecific food competition as related to the specialized foraging adaptations of this species. In the closely related Australian shoveler, the typical situation is for the birds to be in small groups or pairs, widely dispersed (Frith, 1967). Since all the species of shovelers often forage in small groups, with each bird dabbling in the wake of the one in front (Johnsgard, 1965), the maintenance of relatively small flock sizes would be advantageous from this respect as well.

Pair-forming behavior. Pair-forming in wild shovelers begins on the wintering grounds in mid-December and continues until the birds depart for their breeding grounds (McKinney, 1970). The pair-bond is strong and may persist until about hatching or even somewhat afterward. During pair-forming behavior, a variety of male courtship displays are performed (Johnsgard, 1965, see Fig. 24), most of which are derived from motor patterns associated with foraging, such as dabbling, head-dipping, and tipping-up (McKinney, 1970). The primary female display is inciting, and the typical male response to this display is to swim ahead of the inciting female and turn the back of his head toward her. This turning-of-the-back-of-the-head is one of the

Northern shoveler, adult male takeoff

commonest displays observed during pair formation and may persist for a few days or weeks after a pair-bond has been formed (McKinney, 1970). Although unpaired males may attempt to perform the display toward paired females, they never approach the female closely while performing the display.

Copulatory behavior. Copulation is preceded by the usual mutual head-pumping, which is easily distinguished from that associated with aggressive behavior by the lower angle at which the bill is held. Male shovelers may utter a series of soft notes during treading, and immediately after releasing the female they utter a single loud, nasal note followed by a series of repeated wooden sounds while remaining in a rigid posture beside the female, with the body fairly erect and the bill pointed downward (McKinney, 1970).

Nesting and brooding behavior. Females may begin to look for suitable nest sites as early as 27 days before laying begins (McKinney, 1957). Typically, 6 to 8 days are spent in nest construction, and eggs are then laid at the rate of one per day (Girard, 1939). During the egg-laying period, the female may initially spend only an hour or two at the nest but later may be there for the entire morning. The male does not accompany the female to her nest, and she returns to her mate when away from the nest for foraging, resting, or other activities (McKinney, 1957).

During later stages of incubation the female is increasingly reluctant to leave the nest, even when disturbed, and probably remains on it for the last day or so of incubation. About 12 hours elapse between the pipping and hatching of individual eggs, and the female usually leaves the nest within 24 hours of hatching her brood (Girard, 1939). Frequently the male remains with the female and young for a short time after hatching occurs but is not known to participate in brood care.

Postbreeding behavior. Shortly before they begin their flightless period, males may begin to gather in small groups along favored feeding areas. This usually occurs by the end of June in southern Canada, and most of the males are flightless between mid-July and mid-August. Unpaired males might become flightless before those that have bred, and females that have reared families become flightless after rearing their brood, or about the latter part of August (McKinney, 1967).

III. References

General References

The Birds of North America Monographs

Austin, J. E., and M. R. Miller. 1995. Northern Pintail (*Anas acuta*). In *The Birds of North America*, No. 163. (A. Poole and F. Gill, eds.). Philadelphia, PA: The Birds of North America, Inc. 32 pp. (Includes about 200 citations.)

Bielefeld, R. R., M. G. Brasher, and P. N. Gray. 2010. Mottled Duck (*Anas fulvigula*). In *The Birds of North America* online. Edited by A. Poole. Ithaca, NY. This latest version is available at http://bna.birds.cornell.edu

Drilling, N., R. Titman, and F. McKinney. 2002. Mallard (*Anas platyrhynchos*). In *The Birds of North America*, No. 658. (A. Poole and F. Gill, eds.). Philadelphia, PA: The Birds of North America, Inc. 44 pp. (Includes about 225 citations.)

Dubowy, P. J. 1996. Northern Shoveler (*Anas clypeata*). In *The Birds of North America*, No. 217. (A. Poole and F. Gill, eds.). Philadelphia, PA: The Birds of North America, Inc. 24 pp. (Includes about 150 citations.)

Gammonley, J. H. 1996. Cinnamon Teal (*Anas cyanoptera*). In *The Birds of North America*, No. 209. (A. Poole and F. Gill, eds.). Philadelphia, PA: The Birds of North America, Inc. 20 pp. (Includes about 120 citations.)

Hepp, G. R., and F. C. Bellrose. 1995. Wood Duck (*Aix sponsa*). In *The Birds of North America*, No. 169. (A. Poole and F. Gill, eds.). Philadelphia, PA: The Birds of North America, Inc. 24 pp. (Includes about 150 citations.)

Johnson, K. 1995. Green-winged Teal (*Anas carolinensis*). In *The Birds of North America*, No. 193. (A. Poole and F. Gill, eds.). Philadelphia, PA: The Birds of North America, Inc. 20 pp. (Includes about 100 citations.)

Leschack, C. R., S. K. McKnight, and G. R. Hepp. Gadwall (*Anas strepera*). 1997. In *The Birds of North America*, No. 283. (A. Poole and F. Gill, eds.). Philadelphia, PA: The Birds of North America, Inc. 28 pp. (Includes about 175 citations.)

Longcore J. R., D. G. McAuley, G. R. Hepp, and J. M. Rhymer. 2000. American Black Duck (*Anas rubripes*). In *The Birds of North America*, No. 481. (A. Poole and F. Gill, eds.). Philadelphia, PA: The Birds of North America, Inc. 36 pp. (Includes about 250 citations.)

Moorman, T. E., and P. N. Gray. 1994. Mottled Duck (*Anas fulvigula*). In *The Birds of North America*, No. 81. (A. Poole and F. Gill, eds.). Philadelphia, PA: The Birds of North America, Inc. 20 pp. (Includes about 120 citations.)

Mowbray, T. 1999. American Wigeon (*Anas americana*). In *The Birds of North America*, No. 401. (A. Poole and F. Gill, eds.). Philadelphia, PA: The Birds of North America, Inc. 32 pp. (Includes about 150 citations.)

Rohwer, R. C., W. P. Johnson, and E. R. Loos. 2002. Blue-winged Teal (*Anas discors*). In *The Birds of North America*, No. 625. (A. Poole and F. Gill, eds.). Philadelphia, PA: The Birds of North America, Inc. 36 pp. (Includes about 250 citations.)

General Anatidae, Taxonomy, and National Surveys

Aldrich, J. W. 1949. *Migration of Some North American Waterfowl. U.S. Fish and Wildlife Service Special Scientific Report—Wildlife 1*. U.S. Fish and Wildlife Service, Washington, DC.

American Ornithologists' Union. 1957. *Check-list of North American Birds.* 5th ed. American Ornithologists' Union, Washington, DC.

American Ornithologists' Union. 1998. *Check-list of North American Birds.* 7th ed. American Ornithologists' Union, Washington, DC.

Audubon, J. J. 1840–1844. *The Birds of America.* Chevalier, Philadelphia, PA.

Baicich, P. J., and C. J. O. Harrison. 1997. *A Guide to Nests, Eggs, and Nestlings of North American Birds.* 2nd ed. Academic Press Natural World, San Diego, CA.

Baldassarre, G. A. 2014. *Ducks, Geese, and Swans of North America*. Rev. ed. Johns Hopkins University Press. Baltimore, MD. 1,027 pp. (Includes about 3,500 citations.)

Baldassarre, G. A., and E. G. Bolen. 2006. *Waterfowl Ecology and Management*. 2nd ed. Krieger Publishing, Malabar, FL.

Bannerman, D. A. 1958. *Birds of the British Isles*, Vol. 7. Oliver & Boyd, Edinburgh and London, UK.

Batt, B. D. J. (ed.) 1992. *The Ecology and Management of Breeding Waterfowl*. University of Minnesota Press, Minneapolis. 635 pp.

Bauer, K. M., and U. N. Glutz von Blotztheim. 1968–1969. *Handbuch der Vogel Mitteleuropas*. Vols. 2–3. Akademische Verlagsgesellschaft Frankfurt am Main, Germany.

Bellrose, F. 1976. *The Ducks, Geese, and Swans of North America*. 2nd ed. Wildlife Management Institute, Washington, DC.

Bellrose, F. 1980. *The Ducks, Geese, and Swans of North America*. 3rd ed. Wildlife Management Institute, Washington, DC.

Bent, A. C. 1923. *Life Histories of North American Wild Fowl*. Part 1. US National Museum Bulletin 126, US Government Printing Office, Washington, DC.

Bent. A. C. 1925. *Life Histories of North American Wild Fowl*. Part 2. US National Museum Bulletin 130. US Government Printing Office, Washington, DC.

Bezzel, E. 1959. Beiträge zur Biologie der Geschlecter bei Entenvogeln. *Anzeiger der Ornithologischen Gesellschaft in Bayern* 5: 269–355.

Botero, J. E., and D. H. Rusch. 1988. Recoveries of North American waterfowl in the Neotropics. Pp. 469–482 in M. W. Weller, ed. *Waterfowl in Winter*. University of Minnesota Press, Minneapolis.

Canadian Wildlife Service Waterfowl Committee. 2013. *Population Status of Migratory Game Birds in Canada*. Canadian Wildlife Service Migratory Birds Regulatory Report No. 40. Canadian Wildlife Service, Ottawa, ON. http://www.ec.gc.ca/rcommbhr/default.asp?lang=En&n=B2A654BC-1

Carney, S. M. 1964. *Preliminary Key to Age and Sex Identification by Means of Wing Plumage*. US Fish and Wildlife Service Special Scientific Report: Wildlife No. 82.

Cramp, S., and K. E. L. Simmons. 1977. *Handbook of the Birds of Europe, the Middle East, and North Africa: The Birds of the Western Palearctic, Volume 1, Ostrich to Ducks*. Oxford University Press, Oxford, UK.

Delacour, J. 1954–1964. *The Waterfowl of the World*. 4 vol. Country Life, London, UK.

Delacour, J., and E. Mayr. 1945. The family Anatidae. *Wilson Bulletin* 57: 3–55.

del Hoyo, J., A. Elliott, and J. Sargatal. 1992. *Handbook of the Birds of the World. Vol. 1. Ostrich to Ducks*, Lynx Edicions, Barcelona, Spain.

Dementiev, G. P., and N. A. Gladkov, eds. 1967. *Birds of the Soviet Union*. Israel Program for Science Translations, Jerusalem, Israel (translated from Russian).

Feldheim, C. L. 1997. The length of incubation in relation to nest initiation date and clutch size in dabbling ducks. *Condor* 99: 997–1001.

Ferguson, W. H. 1966. Will my birds nest this year? *Modern Game Breeding* 2: 18–20, 34–35.

Gillham, E., and B. Gillham. 1996. *Hybrid Ducks: A Contribution towards an Inventory*. Hythe Printers, Hythe, Kent, UK.

Godfrey, W. E. 1986. *The Birds of Canada*. Rev. ed. (1st ed., 1966). National Museum of Natural Sciences, Ottawa, ON.

Gray, A. P. 1958. *Bird Hybrids: A Check-list with Bibliography*. Technical Communication 13. Commonwealth Agricultural Bureau, Farnham Royal, Bucks, UK.

Havera, S. P. 1999. *Waterfowl of Illinois: Status and Management*. Illinois Natural History Survey, Champaign, IL.

Howell, S. N. G., and S. Webb. 1995. *A Guide to the Birds of Mexico and Northern Central America*. Oxford University Press, Oxford, UK.

Johnsgard, P. A. 1960. Hybridization in the Anatidae and its taxonomic implications. *Condor* 62: 25–33. http://digitalcommons.unl.edu/biosciornithology/71

Johnsgard, P. A. 1961a. Tracheal anatomy of the Anatidae and its taxonomic significance. *Wildfowl Trust Annual Report* 12: 58–69.

Johnsgard, P. A. 1961b. The taxonomy of the Anatidae—A behavioural analysis. *Ibis* 103a: 71–85. http://digitalcommons.unl.edu/johnsgard/29

Johnsgard, P. A. 1965. *Handbook of Waterfowl Behavior*. Cornell University Press, Ithaca, NY. http://digitalcommons.unl.edu/bioscihandwaterfowl/7/

Johnsgard, P. A. 1968. *Waterfowl: Their Biology and Natural History*. University of Nebraska Press, Lincoln. 138 pp.

Johnsgard, P. A. 1975. *Waterfowl of North America*. Indiana University Press, Bloomington. http://digitalcommons.unl.edu/biosciwaterfowlna/1/

Johnsgard, P. A. 1978. *Ducks, Geese, and Swans of the World*. University of Nebraska Press, Lincoln. http://digitalcommons.unl.edu/bioscibucksgeeseswans/

Johnsgard, P. A. 1979a. Anseriformes section (Anatidae and Anhimidae). Pp. 425–506 in E. Mayr, (ed.). *Check-list of the Birds of the World*. Harvard University Press, Cambridge, MA. http://digitalcommons.unl.edu/johnsgard/32

Johnsgard, P. A. 1979b. *A Guide to North American Waterfowl*. Indiana University Press, Bloomington.

Johnsgard, P. A. 1993. *Ducks in the Wild: Conserving Waterfowl and Their Habitats*. Key-Porter, Toronto, ON.

Johnsgard, P. A. 1997. *The Avian Brood Parasites: Deception at the Nest*. Oxford University Press, New York. 409 pp.

Johnsgard, P. A. 2010. *Ducks, Geese, and Swans of the World*. Rev. ed., with supplement: "The World's Waterfowl in the 21st Century." University of Nebraska–Lincoln DigitalCommons and Zea Books. 498 pp. http://digitalcommons.unl.edu/bioscibucksgeeseswans/

Johnsgard, P. A. 2016a. *Swans: Their Biology and Natural History*. University of Nebraska–Lincoln DigitalCommons and Zea Books. 114 pp. http://digitalcommons.unl.edu/zeabook/38/

Johnsgard, P. A. 2016b. *The North American Geese: Their Biology and Behavior*. University of Nebraska–Lincoln DigitalCommons and Zea Books. 159 pp. http://digitalcommons.unl.edu/zeabook/44/

Johnsgard, P. A. 2016c. *The North American Sea Ducks: Their Biology and Behavior*. University of Nebraska–Lincoln DigitalCommons and Zea Books. 256 pp. http://digitalcommons.unl.edu/zeabook/50/

Johnson, K. P. 1997. The evolution of behavior in the dabbling ducks (Anatini): A phylogenetic approach. PhD dissertation, University of Minnesota, Minneapolis.

Kear, J., ed. 2005. *Ducks, Geese, and Swans*. 2 vol. Oxford University Press, Oxford, UK. 898 pp. (Includes more than 3,000 references.)

Kortright, F. H. 1942. *The Ducks, Geese, and Swans of North America*. Wildlife Management Institute, Washington, DC.

Kramer, G. W., E. Carrera, and D. Zaveleta. 1995. Waterfowl harvest and hunter activity in Mexico. *Transactions of the North American Wildlife and Natural Resources Conference* 60: 243–250.

Leopold, S. 1959. *Wildlife of Mexico: The Game Birds and Mammals*. University of California Press, Berkeley.

Linduska, J. P., ed. 1964. *Waterfowl Tomorrow*. US Department of the Interior, Bureau of Sport Fisheries and Wildlife, Washington, DC.

Livezey, B. C. 1997. A phylogenetic classification of waterfowl (Aves: Anseriformes), including selected fossil species. *Annals of the Carnegie Museum* 66: 457–496.

Lorenz, K. 1951–1953. Comparative studies on the behavior of the Anatidae. *Avicultural Magazine* 57: 157–182; 58: 8–17, 61–72, 86–94, 172–184; 59: 24–34, 80–91.

Lutmerding, J. A., and A. S. Love. 2011. *Longevity Records of North American Birds. Version 2011.2*. Bird Banding Laboratory, Patuxent Wildlife Research Center, Laurel, MD.

Madge, S., and H. Burn. 1988. *Waterfowl: An Identification Guide to the Ducks, Geese, and Swans of the World*. Houghton Mifflin, Boston.

Martin, A. C., H. S. Zim, and A. L. Nelson. 1951. *American Wildlife and Plants*. McGraw-Hill, New York.

McKinney, F. 1965a. Spacing and chasing in breeding ducks. *Wildfowl Trust Annual Report* 16: 92–106.

Mosby, H. S., ed. 1967. *Wildlife Investigational Techniques*. 2nd ed. Wildlife Society, Washington, DC.

Munroe, B. L. J. 1968. *A Distributional Survey of the Birds of Honduras*. Ornithological Monographs, American Ornithologists' Union. No. 7. 758 pp.

Nelson, A. D., and A. C. Martin. 1953. Gamebird weights. *Journal of Wildlife Management* 17: 36–42.

Ogilvie, M. A. 1975. *Ducks of Britain and Europe*. T. & A. D. Poyser, Berkhamsted, UK.

Owen, M. 1977. *Wildfowl of Europe*. Macmillan, London, UK.

Owen, M., G. L. Atkinson-Willes, and D. Salmon. 1986. *Wildfowl in Great Britain*. 2nd ed. Cambridge University Press, Cambridge, UK.

Palmer, R. S., ed. 1976. *Handbook of North American Birds*, Vol. 2: Waterfowl, Part 1. Yale University Press, New Haven, CT.

Phillips, J. C. 1922–1926. *A Natural History of the Ducks*. 4 vol. Houghton Mifflin, Boston.

Raftovich, R. V., S. C. Chandler, and K. A. Wilkins. 2015. *Migratory Bird Hunting Activity and Harvest during the 2013–14 and 2014–15 Hunting Seasons*. US Fish and Wildlife Service, Laurel, MD.

Rodner, C., M. Lentino, and R. Restall. 2000. *Checklist of the Birds of Northern South America*. Yale University Press, New Haven, CT.

Salomonsen, F. 1950. *The Birds of Greenland*. Part 1. Ejnar Munksgaard, Copenhagen, Denmark.

Sauer, J. R, , W. A. Link, J. E. Fallon, K. L. Pardieck, and D. J. Ziolkowski, Jr. 2013. *The North American Breeding Bird Survey 1966–2011: Summary Analysis and Species Accounts*. North American Fauna Number 79: 1–3.

Saunders, G. B., and D. C. Saunders. 1981. *Waterfowl and Their Wintering Grounds in Mexico, 1937–1964*. Resource Publication 138. US Fish and Wildlife Service, Washington, DC.

Schiøler, E. 1925–1926. *Danmarks Fugle*. 2 vol. Gyldendelske, Denmark.

Scott, P., and H. Boyd. 1957. *Wildfowl of the British Isles*. Country Life Ltd, London, UK.

Smith, A. R. 1996. *Atlas of Saskatchewan Birds*. Saskatchewan Natural History Society, Regina, SK.

Snyder, L. L. 1957. *Arctic Birds of Canada*. University of Toronto Press, Toronto, ON.

Stevenson, H. M., and B. H. Anderson. 1994. *The Birdlife of Florida*. University Press of Florida, Gainesville.

Stiles, F. G., and A. F. Skutch. 1989. *A Guide to the Birds of Costa Rica*. Cornell University Press, Ithaca, NY.

Todd, F. S. 1979. *Waterfowl: Ducks, Geese, and Swans of the World*. Harcourt Brace Jovanovich, New York, and Sea World Press, San Diego, CA.

Todd, F. S. 1996. *Natural History of the Waterfowl*. Ibis Publishing, Vista, CA.

US Fish and Wildlife Service [USFWS]. 2009. *Migratory Bird Hunting Activity and Harvest during the 2007 and 2008 Hunting Seasons*. US Fish and Wildlife Service, Laurel, MD.

US Fish and Wildlife Service. 2009. *Waterfowl Population Status, 2009*. US Fish and Wildlife Service, Laurel, MD.

US Fish and Wildlife Service. 2013. *North American Breeding Bird Survey, Results and Analysis 1966–2013*. Patuxent Wildlife Research Center, Laurel, MD. *Analysis:* http://www.pwrc.usgs.gov/bbs.

US Fish and Wildlife Service. 2014. *Waterfowl Population Status, 2014*. US Fish and Wildlife Service, Laurel, MD.

US Fish and Wildlife Service. 2015. *Waterfowl Population Status, 2015*. US Fish and Wildlife Service, Laurel, MD.

US Fish and Wildlife Service. 2016. *Waterfowl Population Status, 2016*. US Fish and Wildlife Service, Laurel, MD.

Weller, M. W., ed. 1998. *Waterfowl in Winter*. University of Minnesota Press, Minneapolis.

Zimpfer, N. L., W. E. Rhodes, E. D. Silverman, G. S. Zimmerman, and K. D. Richkus. 2014. *Trends in Duck Breeding Populations, 1955–2014*. Administrative Report. US Fish and Wildlife Service, Laurel, MD.

State and Regional Bird Surveys

Adamus, P. R., K. Larsen, G. Gillson, and C. R. Miller. 2001. *Oregon Breeding Bird Atlas*. Oregon Field Ornithologists, Eugene.

Alcorn, J. R. 1988. *The Birds of Nevada*. Fairview West Publishing, Fallon, NV.

Andrews, R., and R. Righter. 1992. *Colorado Birds*. Denver Museum of Natural History, Denver, CO.

Bailey, A. M. and R. J. Niedrach. 1965. *Birds of Colorado*. Denver Museum of Natural History, Denver, CO.

Bailey, F. M. 1928. *Birds of New Mexico*. New Mexico Department of Fish and Game, Santa Fe.

Baumgartner, F. M., and A. M. Baumgartner. 1992. *Oklahoma Bird Life*. University of Oklahoma Press, Norman.

Benson, K. L. P., and K. A. Arnold. 2001. *The Texas Breeding Bird Atlas*. Texas A&M University System, College Station and Corpus Christi. http://txtbba.tamu.edu

Brown, D. E. 1985. *Arizona Wetlands and Waterfowl*. University of Arizona Press, Tucson.

Brown, M. B., and P. A. Johnsgard. 2013. *Birds of the Central Platte River Valley and Adjacent Counties*. University of Nebraska–Lincoln DigitalCommons and Zea Books. 182 pp. http://digitalcommons.unl.edu/zeabook/15/

Burleigh, T. D. 1944. *The Bird Life of the Gulf Coast Region of Mississippi.* Louisiana State University, Museum of Zoology Occasional Papers No. 20: 329–490.

Burleigh, T. D. 1972. *Birds of Idaho.* Caxton Printers, Caldwell, ID.

Byrd, G. V., D. L. Johnson, and D. D. Gibson. 1974. The birds of Adak Island, Alaska. *Condor* 76: 288–300.

Cadman, M. D., P. J. F. Engels, and F. M. Helleiner, eds. 2016. *Atlas of the Breeding Birds of Ontario.* 2nd ed. Federation of Ontario Naturalists and Long Point Bird Observatory, Bird Studies Canada, Environment Canada, Ontario Field Ornithologists, Ontario Ministry of Natural Resources, and Ontario Nature, Toronto, ON.

Campbell, R. W., N. K. Dawe, I. McTaggart-Cowan, J. M. Cooper, G. W. Kaiser, and M. C. E. McNall. 1990. *The Birds of British Columbia. Volume 1: Nonpasserines, Introduction and Loons through Waterfowl.* University of British Columbia Press, Vancouver.

Canterbury, J., P. A. Johnsgard, and H. Downing. 2013. *Birds and Birding in Wyoming's Bighorn Mountains Region.* University of Nebraska–Lincoln DigitalCommons and Zea Books. 260 pp. http://digitalcommons.unl.edu/zeabook/18/

Castrale, J. S., E. M. Hopkins, and C. E. Keller. 1998. *Atlas of Breeding Birds of Indiana.* Indiana Department of Natural Resources, Indianapolis.

Cheng, Tso-hsin. 1963. *China's Economic Fauna: Birds.* Science Publishing Society, Peiping (Beijing). (Translated by Joint Publication Research Service, Washington, DC, 1964.)

Cogswell, H. L. 1977. *Water Birds of California.* University of California Press, Berkeley.

Corman, T. E., and C. Wise-Gervais. 2005. *Arizona Breeding Bird Atlas.* University of New Mexico Press, Albuquerque.

Courcelles, R., and J. Bédard. 1979. Habitat selection by dabbling ducks in the Baie Noire Marsh, southwestern Québec. *Canadian Journal of Zoology* 57: 2230–2238.

Davidson, P. J. A., R. J. Cannings, A. R. Couturier, D. Lepage, and C. M. D. Corrado, eds. 2008 et seq. *The Atlas of the Breeding Birds of British Columbia.* Bird Studies Canada. Delta, BC. Species accounts are available at http://www.birdatlas.bc.ca/accounts/toc.jsp?show=species

Dinsmore, J. J., T. H. Kent, D. Koenig, P. C. Petersen, and D. M. Roosa. 1984. *Iowa Birds.* Iowa State University Press, Ames.

Erskine, A. J. 2016 (1st ed., 1992). *Atlas of Breeding Birds of the Maritime Provinces.* Bird Studies Canada, Sackville, NS.

Gabrielson, I. N., and F. C. Lincoln. 1959. *Birds of Alaska.* Stackpole, Harrisburg, PA, and Wildlife Management Institute, Washington, DC.

Gauthier, J., and Y. Aubry, eds. 1996. *The Breeding Birds of Québec: Atlas of the Breeding Birds of Southern Québec.* Province of Québec Society for the Protection of Birds, Canadian Wildlife Service, Québec Region, Montréal.

Gibson, D. D., and G. V. Byrd. 2007. *Birds of the Aleutian Islands, Alaska.* Nuttall Ornithological Club, Cambridge, MA, and American Ornithologists' Union, Washington, DC.

Gill, R. E., Jr., M. R. Petersen, and P. D. Jorgensen. 1981. Birds of the north-central Alaska Peninsula, 1976–1980. *Arctic* 34: 286–306.

Gilligan, J., M. Smith, D. Rogers, and A. Contreras, eds. 1994. *Birds of Oregon: Status and Distribution.* Cinclus Publications, McMinnville, OR.

Granlund, J., G. A. McPeek, R. J. Adams, and C. Callog. 1994. *The Birds of Michigan.* Indiana University Press, Bloomington.

Grinnell, J., and A. H. Miller. 1944. *The Distribution of the Birds of California.* Pacific Coast Avifauna, Number 27. Cooper Ornithological Club, Berkeley, CA.

Griscom, L., and D. E. Snyder. 1955. *The Birds of Massachusetts: An Annotated and Revised Check List.* Peabody Museum, Salem, MA.

Gromme, O. J. 1963. *Birds of Wisconsin.* University of Wisconsin Press, Madison.

Hess, G. K., R. L. West, M. V. Barnhill, III, and L. M. Fleming. 2000. *Birds of Delaware.* University of Pittsburgh Press, Pittsburgh, PA.

Jackson, L. S., C. A. Thompson, and J. J. Dinsmore. 1996. *The Iowa Breeding Bird Atlas.* University of Iowa Press, Iowa City.

Janssen, R. B. 1987. *Birds in Minnesota.* University of Minnesota Press, Minneapolis.

Johnsgard, P. A. 1953. Effects of water fluctuation and vegetation change on bird populations, especially waterfowl. MS thesis, Washington State College, Pullman.

Johnsgard, P. A. 1986. *Birds of the Rocky Mountains, with Particular Reference to National Parks in the Northern Rocky Mountain Region.* Colorado Associated University Press, Boulder. 504 pp.

Johnsgard, P. A. 2009. *Birds of the Great Plains: Breeding Species and Their Distribution.* University of Nebraska Press, Lincoln. Revised edition with a 2009 supplement at http://digitalcommons.unl.edu/bioscibirdsgreatplains/1/

Johnsgard, P. A. 2012a. *Wetland Birds of the Central Plains: South Dakota, Nebraska, and Kansas.* University of Nebraska–Lincoln DigitalCommons and Zea Books. 275 pp. http://digitalcommons.unl.edu/zeabook/8/

Johnsgard, P. A. 2012b. *Wings over the Great Plains: The Central Flyway.* University of Nebraska–Lincoln DigitalCommons and Zea Books. 249 pp. http://digitalcommons.unl.edu/zeabook/13/

Johnsgard, P. A. 2013. *The Birds of Nebraska.* Rev. ed. University of Nebraska–Lincoln DigitalCommons and Zea Books. 150 pp. http://digitalcommons.unl.edu/zeabook/17/

Johnsgard, P. A. 2015. *Global Warming and Population Responses among Great Plains Birds.* University of Nebraska–Lincoln DigitalCommons and Zea Books. 384 pp. http://digitalcommons.unl.edu/zeabook/26/

Johnsgard, P. A., and I. O. Buss. 1956. Waterfowl sex ratios during spring in Washington State and their interpretation. *Journal of Wildlife Management* 20: 384–388.

Johnson, W. P., and M. W. Lockwood. 2013. *Texas Waterfowl.* Texas A&M University Press, College Station. 173 pp.

Johnston, R. F. 1964. The breeding birds of Kansas. Lawrence, KS: *University of Kansas, Publications of the Museum of Natural History* 12: 57–655.

Kessel, B. 1989. *Birds of the Seward Peninsula, Alaska: Their Biogeography, Seasonality, and Natural History.* University of Alaska Press, Fairbanks.

Kessel, B., and D. G. Gibson. 1978. *Status and Distribution of Alaska Birds. Studies in Avian Biology 1.* Cooper Ornithological Society, Los Angeles, CA.

Kingery, H. E., ed. 1997. *Colorado Breeding Bird Atlas.* Colorado Bird Partnership, Denver.

Larrison, E. J., and K. G. Sonnenburg. 1968. *Washington Birds: Their Location and Identification.* Seattle Audubon Society, Seattle, WA.

Lockwood, M. W., and B. Freeman. 2014. *Handbook of Texas Birds.* 2nd ed. Texas A&M University Press, College Station.

Marks, J., P. Hendricks, and D. Casey. 2016. *Birds of Montana.* Buteo Books, Arlington, VA. 672 pp.

McGowan, K. J., and K. Corwin. 2008. *The Second Atlas of Breeding Birds in New York State.* Cornell University Press, Ithaca, NY.

McPeek, G. A., ed. 1994. *The Birds of Michigan.* Indiana University Press, Bloomington.

McWilliams, G. M., and D. W. Brauning. 2000. *The Birds of Pennsylvania.* Cornell University Press, Ithaca, NY.

Molhoff, W. J. 2016. *The Second Nebraska Breeding Bird Atlas.* Bulletin of the University of Nebraska State Museum 29. 304 pp.

Murie, O. J. 1959. *Fauna of the Aleutian Islands and Alaska Peninsula.* US Department of the Interior, Fish and Wildlife Service, North American Fauna, No. 61: 1–406.

Oakleaf, B., B. Luce, S. Ritter, and A. Cerovski, eds. 1992. *Wyoming Bird and Mammal Atlas.* Wyoming Game and Fish Department, Lander.

Oberholser, H. C. 1974. *The Bird Life of Texas.* Vol. 1. University of Texas Press, Austin.

Palmer, R. S. 1949. *Maine Birds.* Bulletin of the Museum of Comparative Zoology 102. Museum of Comparative Zoology, Cambridge, MA.

Parmelee, D. F., and H. A. Stephens, and R. H. Schmidt. 1967. The birds of southeastern Victoria Island and adjacent small islands. *National Museums of Canada Bulletin* 222: 1–229.

Parmelee, D. F., and S. D. MacDonald. 1960. The birds of west-central Ellesmere Island and adjacent areas. *National Museums of Canada Bulletin* 169: 1–101.

Peterjohn, B. G. 1989. *The Birds of Ohio.* Indiana University Press, Bloomington.

Peterjohn, B. G., and D. L. Rice. 1991. *The Ohio Breeding Bird Atlas.* Ohio Department of Natural Resources, Columbus.

Peters, H. S., and T. D. Burleigh. 1951. *The Birds of Newfoundland.* Newfoundland Department of Natural Resources, St. Johns.

Peterson, R. A. 1995. *The South Dakota Breeding Bird Atlas.* South Dakota Ornithologists' Union, Aberdeen, SD.

Phillips, A. R., J. Marshall, and G. Monson. 1964. *Birds of Arizona.* University of Arizona Press, Tucson.

Pranty, B., K. A. Radamaker, and G. Kennedy. 2006. *Birds of Florida*. Lone Pine Publishing International, Auburn, WA.

Robbins, M. B., and D. A. Easterla. 1992. *Birds of Missouri: Their Distribution and Abundance*. University of Missouri Press, Columbia.

Robbins, S. D., Jr. 1991. *Wisconsin Birdlife*. University of Wisconsin Press, Madison.

Roberts, T. S. 1932. *The Birds of Minnesota*. 2 vols. University of Minnesota Press, Minneapolis.

Root, T. 1988. *Atlas of Wintering North American Birds: An Analysis of Christmas Bird Count Data*. University of Chicago Press, Chicago.

Rosenberg, K. V., R. D. Ohmart, W. C. Hunter, and B. W. Anderson. 1991. *Birds of the Lower Colorado River Valley*. University of Arizona Press, Tucson.

Semenchuk, G. P. 1992. *The Atlas of Breeding Birds of Alberta*. Federation of Alberta Naturalists, Edmonton.

Small, A. 1994. *California Birds: Their Status and Distribution*. Ibis Publishing, Temecula, CA.

Smith, A. R. 1996. *Atlas of Saskatchewan Birds*. Saskatchewan Natural History Society, Regina.

Sprunt, A., Jr., and E. B. Chamberlain. 1949. *South Carolina Bird Life*. University of South Carolina Press, Columbia.

Stevenson, H. M., and B. H. Anderson. 1994. *The Birdlife of Florida*. University Press of Florida, Gainesville.

Stewart, R. E. 1975. *Breeding Birds of North Dakota*. Tri-College Center for Regional Studies, Fargo, ND.

Sutton, G. M. 1967. *Oklahoma Birds*. University of Oklahoma Press, Norman.

Szymczak, M. R. 1986. *Characteristics of Duck Populations in the Intermountain Parks of Colorado*. Technical Publication 35. Colorado Division of Wildlife, Denver.

Texas Game, Fish, and Oyster Commission. 1945. *Principal Game Birds and Mammals of Texas*. Texas Game, Fish, and Oyster Commission, Austin.

Thompson, M. C., C. A. Ely, B. Gress, C. Otte, S. T. Pati, D. Seibl, and E. A. Young. 2011. *Birds of Kansas*. University Press of Kansas, Lawrence.

Todd, W. E. C. 1963. *Birds of the Labrador Peninsula and Adjacent Areas*. University of Toronto Press, Toronto.

Tufts, R. W. 1986. *Birds of Nova Scotia*. 3rd ed. Nimbus Publishing and Nova Scotia Museum, Halifax, NS.

Veit, R. R., and W. R. Petersen. 1993. *Birds of Massachusetts*. Massachusetts Audubon Society, Lincoln, MA.

Wahl, T. R., B. Tweit, and S. G. Mlodinow. 2005. *Birds of Washington: Status and Distribution*. Oregon State University Press, Corvallis.

Wetlands International. 2006. *Waterbird Population Estimates*. 4th ed. Wetlands International, Wageningen, Netherlands.

Wetlands International. 2012. *Waterbird Population Estimates*. 5th ed. Wetlands International, Wageningen, Netherlands.

Zeranski, J. D., and T. R. Baptist. 1990. *Connecticut Birds*. University Press of New England, Hanover, NH.

Zimmerman, D. A., and J. Van Tyne. 1959. A distributional check-list of the birds of Michigan. *University of Michigan Museum of Zoology Occasional Papers*, No. 608: 1–63.

Multiple Taxa Studies

Note: "Multiple" here indicates that three or more duck species are evidently involved in the study. Those studies involving only two named species are listed in the reference subsections of each of the two species.

Addy, C. E. Atlantic Flyway. Pp. 167–184 in J. P. Linduska, ed. *Waterfowl Tomorrow*. US Department of the Interior, Bureau of Sport Fisheries and Wildlife, Washington, DC.

Aldrich, J. W. 1949. *Migration of Some North American Waterfowl. US Fish and Wildlife Service Special Scientific Report—Wildlife 1*. US Fish and Wildlife Service, Washington, DC.

Anderson, W. 1957. A waterfowl nesting study in the Sacramento Valley, California, 1955. *California Fish and Game* 43: 71–90.

Anderson, W. 1960. A study of waterfowl nesting in the Suisun marshes. *California Fish and Game* 46: 217–226.

Anderson, W. 1965. Waterfowl production in the vicinity of gull colonies. *California Fish and Game* 51: 5–15.

Bartonek, J. C. 1972. Summer foods of American wigeon, mallards, and a green-winged teal near Great Slave Lake, N.W.T. *Canadian Field-Naturalist* 86: 373–376.

Blums, P., A. Mednis, I. Bauga, J. D. Nichols, and J. E. Hines. 1996. Age-specific survival and philopatry in three species of European ducks: a long-term study. *Condor* 98: 61–74.

Brown, D. E. 1982. Sex ratios, sexual selection, and sexual dimorphism in waterfowl. *American Birds* 36: 258–260.

Chamberlain, E. B. 1960. *Florida Waterfowl Populations, Habitats, and Managements.* Florida Game and Fresh Water Fish Commission, Technical Bulletin, No.7, pp. 1–62.

Clark, R. C., L. G. Sugden, R. K. Brace, and D. J. Nieman. 1988. The relationship between nesting chronology and vulnerability to hunting of dabbling ducks. *Wildfowl* 39: 137–144.

Coulter, M. W. 1955. Spring food habits of surface-feeding ducks in Maine. *Journal of Wildlife Management* 19: 263–267.

Coulter, M. W. 1957. Predation by snapping turtles upon aquatic birds in Maine marshes. *Journal of Wildlife Management* 21: 17–21.

Courcelles, R., and J. Bédard. 1979. Habitat selection by dabbling ducks in the Baie Noire Marsh, southwestern Québec. *Canadian Journal of Zoology* 57: 2230–2238.

Duebbert, H. F., and A. M. Frank. 1984. Value of prairie wetlands to duck broods. *Wildlife Society Bulletin* 12: 27–34.

Duebbert, H. F., and J. T. Lokemoen. 1976. Duck nesting in fields of undisturbed grass-legume cover. *Journal of Wildlife Management* 40: 39–49.

Duebbert, H. F., J. T. Lokemoen, and D. E. Sharp. 1986. Nest sites of ducks in grazed mixed-grass prairie in North Dakota. *Prairie Naturalist* 18: 99–108.

Evans, C. D., A. S. Hawkins, and W. H. Marshall. 1952. *Movements of Waterfowl Broods in Manitoba. Special Scientific Report—Wildlife 16.* US Fish and Wildlife Service, Washington, DC.

Evans, C. D., and K. E. Black. 1956. *Duck Production Studies on the Prairie Potholes of South Dakota. Special Scientific Report—Wildlife 32.* US Fish and Wildlife Service, Washington, DC.

Feldheim, C. L. 1997. The length of incubation in relation to nest initiation date and clutch size in dabbling ducks. *Condor* 99: 997–1001.

Fredrickson, L. H., and M. E. Heitmeyer. 1988. Waterfowl use of forested wetlands of the southern United States: An overview. Pp. 307–323 in M. W. Weller, ed. *Waterfowl in Winter.* University of Minnesota Press, Minneapolis.

Gammonley, J. H. 1996. Seasonal use of montane wetlands by waterfowl on the rim of the Colorado Plateau. PhD dissertation, University of Missouri, Columbia.

Havera, S. P. 1999. *Waterfowl of Illinois: Status and Management.* Special Publication 21. Illinois Natural History Survey, Urbana.

Higgins, K. F., T. A. Klett, and H. W. Miller. 1992. *Waterfowl Production on the Woodworth Station in South-Central North Dakota, 1965–1981.* Resource Publication 180. US Fish and Wildlife Service, Washington, DC.

Hochbaum, H. A. 1944. *The Canvasback on a Prairie Marsh.* Wildlife Management Institute, Washington, DC, and Stackpole, Harrisburg, PA.

Johnson, D. H., J. D. Nichols, and M.D. Schwartz. 1992. Population dynamics of breeding waterfowl. Pp. 446–485 in B. Batt et al., eds. *The Ecology and Management of Breeding Waterfowl.* University of Minnesota Press, Minneapolis.

Kear, J., and P. A. Johnsgard. 1968. Foraging dives by surface-feeding ducks. *Wilson Bulletin* 80: 231.

Keith, L. B. 1961. *A Study of Waterfowl Ecology on Small Impoundments in Southeastern Alberta.* Wildlife Monographs, No. 6. The Wildlife Society, Bethesda, MD.

Krapu, G. L. 1979. Nutrition of female dabbling ducks during reproduction. Pp. 59–70 in T. A. Bookhout, ed. *Waterfowl and Wetlands: An Integrated Review.* La Crosse Printing, La Crosse, WI.

Krapu, G. L. 2000. Temporal flexibility of reproduction in temperate-breeding dabbling ducks. *Auk* 117: 640–650.

Krapu, G. L., P. J. Pietz, D. A. Brandt, and R. R. Cox, Jr. 2004b. Does presence of permanent fresh water affect recruitment in prairie-nesting dabbling ducks? *Journal of Wildlife Management* 68: 332–341.

Kruse, A. D., and B. S. Bowen. 1996. Effects of grazing and burning on densities and habitats of breeding ducks in North Dakota. *Journal of Wildlife Management* 60: 233–246.

Lee, F. B., R. L. Jessen, N. J. Ordal, R. I. Benson, J. P. Lindmeier, and L. L. Johnson. 1964a. *Waterfowl in Minnesota*. Technical Bulletin 7. Minnesota Department of Conservation, St. Paul.

Lee, F. B., R. L. Jessen, N. J. Ordal, R. I. Benson, J. P. Lindmeier, and L. L. Johnson. 1964b. *Ducks and Land Use in Minnesota: Four Studies*. Minnesota Department of Conservation Technical Bulletin, No. 8. pp. 1–140.

Leitch, W. G., and R. M. Kaminski. 1985. Long-term wetland-waterfowl trends in Saskatchewan grasslands. *Journal of Wildlife Management* 49: 212–222.

Lokemoen, J. T. 1991. Brood parasitism among waterfowl nesting on islands and peninsulas in North Dakota. *Condor* 93: 340–345.

Lokemoen, J. T., and R. O. Woodward. 1992. Nesting waterfowl and water birds on natural islands in the Dakotas and Montana. *Wildlife Society Bulletin* 20: 163–171.

Lokemoen, J. T., D. H. Johnson, and D. E. Sharp. 1990b. Weights of wild mallard *Anas platyrhynchos*, gadwall *A. strepera*, and blue-winged teal *A. discors* during the breeding season. *Wildfowl* 41: 122–130.

Lokemoen, J. T., H. F. Duebbert, and D. E. Sharp. 1984. Nest spacing, habitat selection, and behavior of waterfowl on Miller Lake Island, North Dakota. *Journal of Wildlife Management* 48: 309–321.

Lokemoen, J. T., H. F. Duebbert, and D. E. Sharp. 1990a. *Homing and Reproductive Habits of Mallards, Gadwalls, and Blue-winged Teal*. Wildlife Monographs 106. The Wildlife Society, Bethesda, MD.

Loos, E. R., and F. C. Rohwer. 2004. Laying-stage nest attendance and onset of incubation in prairie nesting ducks. *Auk* 121: 587–599.

McKinney, F. 1970. The displays of four species of blue-winged ducks. *Living Bird* 9: 29–64.

McKinney, F. 1975. The evolution of duck displays. Pp. 331–357 in G. Baerends, C. Beer, and A. Manning, eds. *Function and Evolution in Behaviour: Essays in Honour of Professor Niko Tinbergen*. Clarendon Press, Oxford, UK.

McMahan, C. A. 1970. Food habits of ducks wintering on Laguna Madre, Texas. *Journal of Wildlife Management* 34: 946–949.

Migoya, R. 1989. Chronology of migration, harvest, and food habits of wintering waterfowl in Sinaloa, México. MS thesis, Auburn University, Auburn, AL.

Migoya, R., and G. A. Baldassarre. 1993. Harvest and food habits of waterfowl wintering in Sinaloa, México. *Southwestern Naturalist* 38: 168–171.

Miller, M. R., J. Beam, and D. P. Connelly. 1988. Dabbling duck harvest dynamics in the Central Valley of California: implications for recruitment. Pp. 553–569 in M. W. Weller, ed. *Waterfowl in Winter*. University of Minnesota Press, Minneapolis.

Monda, M. J., and J. T. Ratti. 1988. Niche overlap and habitat use by sympatric duck broods in eastern Washington. *Journal of Wildlife Management* 52: 95–103.

Moyle, J. B., F. B. Lee, R. L. Jessen, N. J. Ordal, R. I. Benson, J. P. Lindmeier, and L. L. Johnson. 1964. *Waterfowl in Minnesota*. Technical Bulletin 7. Division of Game and Fish, Minnesota Department of Conservation, St. Paul.

Murdy, H. W. 1964. *Population Dynamics and Breeding Biology of Waterfowl on the Yellowknife Study Area, Northwest Territories: Annual Progress Report*. Northern Prairie Wildlife Research Center, US Fish and Wildlife Service, Jamestown, ND.

Murdy, H. W. 1965. *Population Dynamics and Breeding Biology of Waterfowl on the Yellowknife Study Area, Northwest Territories: Annual Progress Report*. Northern Prairie Wildlife Research Center, US Fish and Wildlife Service, Jamestown, ND.

Musgrove, J. W., and M. R. Musgrove. 1947. *Waterfowl in Iowa*. 2nd ed. State Conservation Commission, Des Moines, IA.

Myers, T. L. 1982. Ecology of nesting waterfowl on Anderson Mesa in north central Arizona. MS thesis, University of Arizona, Tucson.

Nudds, T. D., and J. N. Bowlby. 1984. Predator-prey size relationships in North American dabbling ducks. *Canadian Journal of Zoology* 62: 2002–2008.

Nudds, T. D., and R. W. Cole. 1991. Changes in populations and breeding success of boreal forest ducks. *Journal of Wildlife Management* 55: 569–573.

Oring, L. W. 1964. Behavior and ecology of certain ducks during the postbreeding period. *Journal of Wildlife Management* 28: 223–233.

Petrula, M. J. 1994. Nesting ecology of ducks in interior Alaska. MS thesis, University of Alaska–Fairbanks, Fairbanks.

Pierluissi, S., S. L. King, and M. D. Kaller. 2010. Waterbird nest density and nest survival in rice fields of southwestern Louisiana. *Waterbirds* 33: 323–330.

Pieron, M. R., and F. C. Rohwer. 2010. Effects of large-scale predator reduction on nest success of upland nesting ducks. *Journal of Wildlife Management* 74: 124–132.

Pöysä, H. 1985. Circumstantial evidence of foraging interference of dabbling ducks. *Wilson Bulletin* 97: 541–543.

Pöysä, H. 1986. Species composition and size of dabbling duck (*Anas* spp.) feeding groups: are foraging interactions important determinants? *Ornis Fennica* 63: 33–41.

Renouf, R. N. 1972. Waterfowl utilization of beaver ponds in New Brunswick. *Journal of Wildlife Management* 36: 740–744.

Reynolds, R. E. 2005. The Conservation Reserve Program and duck production in the United States' Prairie Pothole region. Pp. 144–148 in A. W. Allen and M. W. Vandever, eds. *The Conservation Reserve Program: Planting for the Future; Proceedings of a National Conference. Scientific Investigations Report 2005-5145.* US Geological Survey, Reston, VA.

Reynolds, R. E., T. L. Shaffer, C. R. Loesch, and R. R. Cox, Jr. 2006. The Farm Bill and duck production in the Prairie Pothole region: increasing the benefits. *Wildlife Society Bulletin* 34: 963–974.

Reynolds, R. E., T. L. Shaffer, R. W. Renner, W. E. Newton, and B. D. J. Batt. 2001. Impact of the Conservation Reserve Program on duck recruitment in the US Prairie Pothole region. *Journal of Wildlife Management* 65: 765–780.

Rienecker, W. C., and W. Anderson. 1960. A waterfowl nesting study on Tule Lake and Lower Klamath National Wildlife Refuges, 1957. *California Fish and Game* 46: 481–506.

Rohwer, F. C., and S. Freeman. 1989. The distribution of conspecific nest parasitism in birds. *Canadian Journal of Zoology* 67: 239–253.

Ross, R. K., K. F. Abraham, T. R. Gadawski, R. S. Rempel, T. S. Gabor, and R. Maher. 2002. Abundance and distribution of breeding waterfowl in the Great Clay Belt of northern Ontario. *Canadian Field-Naturalist* 116: 42–50.

Ruwaldt, J. J., L. D. Flake, and J. M. Gates. 1979. Waterfowl pair use of natural and man-made wetlands in South Dakota. *Journal of Wildlife Management* 43: 375–383.

Sargeant, A. B., and D. G. Raveling. 1992. Mortality during the breeding season. Pp. 396–422 in B. D. J. Batt, A. D. Afton, M. G. Anderson, C. D. Ankney, D. H. Johnson, J. A. Kadlec, and G. L. Krapu, eds. *Ecology and Management of Breeding Waterfowl.* University of Minnesota Press, Minneapolis.

Sargeant, A. B., R. J. Greenwood, M. A. Sovada, and T. L. Shaffer. 1993. *Distribution and Abundance of Predators That Affect Duck Production—Prairie Pothole Region.* Resource Publication 194. US Fish and Wildlife Service, Washington, DC.

Sargeant, A. B., S. H. Allen, and R. T. Eberhardt. 1984. *Red Fox Predation on Breeding Ducks in Midcontinent North America.* Wildlife Monographs 89. The Wildlife Society, Washington, DC.

Smith, A. G. 1971. *Ecological Factors Affecting Waterfowl Production in the Alberta Parklands.* Bureau of Sport Fisheries and Wildlife, Fish and Wildlife Service, Washington, DC, Research Publication 98.

Sovada, M. A., R. M. Anthony, and B. D. J. Batt. 2001. Predation on waterfowl in arctic tundra and prairie breeding areas: A review. *Wildlife Society Bulletin* 29: 6–15.

Stafford, J. D., R. M. Kaminski, K. J. Reinecke, and S. W. Manley. 2006. Waste rice for waterfowl in the Mississippi Alluvial Valley. *Journal of Wildlife Management* 70: 61–69.

Stewart, R. E. 1962. *Waterfowl Populations in the Upper Chesapeake Region.* US Department of the Interior, Fish and Wildlife Service, Special Scientific Report: Wildlife, No. 65, pp. 1–208.

Stewart, R. E., and H. A. Kantrud. 1973. Ecological distribution of breeding waterfowl populations in North Dakota. *Journal of Wildlife Management* 37: 39–50.

Stewart, R. E., and H. A. Kantrud. 1974. Breeding waterfowl populations in the Prairie Pothole Region of North Dakota. *Condor* 76: 70–79.

Stutzenbaker, C. D., and M. W. Weller. 1989. The Texas coast. Pp. 385–405 in L. M. Smith, R. L. Pederson, and R. M. Kaminski, eds. *Habitat Management for Migrating and Wintering Waterfowl in North America.* Texas Tech University Press, Lubbock.

Sugden, L. G. 1973. *Feeding Ecology of Pintail, Gadwall, American Widgeon, and Lesser Scaup Ducklings in Southern Alberta.* Report Series 24. Canadian Wildlife Service, Ottawa, ON.

Swanson, G. A., G. L. Krapu, and J. R. Serie. 1979. Foods of laying female dabbling ducks on the breeding grounds. Pp. 47–57 in T. A. Bookhout, ed. *Waterfowl and Wetlands: An Integrated Review.* Northcentral Section of the Wildlife Society, Madison, WI.

Szymczak, M. R. 1986. *Characteristics of Duck Populations in the Intermountain Parks of Colorado.* Technical Publication 35. Colorado Division of Wildlife, Denver.

Thompson, J. D., and G. A. Baldassarre. 1991. Activity patterns of Nearctic dabbling ducks wintering in Yucatán, México. *Auk* 108: 934–941.

Thompson, J. D., and G. A. Baldassarre. 1992. Dominance relationships of dabbling ducks wintering in Yucatán, México. *Wilson Bulletin* 104: 529–536.

Thompson, J. D., B. J. Sheffer, and G. A. Baldassarre. 1992. Food habits of selected dabbling ducks wintering in Yucatán, México. *Journal of Wildlife Management* 56: 740–744.

Thorpe, P., R. Drewien, and G. O. Garcia. 2006. *Winter Waterfowl Survey: Mexico Interior Highlands and Lower West Coast, January 2006.* US Fish and Wildlife Service, Washington, DC.

Titman, R. D., and N. R. Seymour. 1981. A comparison of pursuit flights by six North American ducks of the genus *Anas. Wildfowl* 32: 11–18.

Toft, C. A., D. L. Trauger, and H. W. Murdy. 1984. Seasonal decline in brood sizes of sympatric waterfowl (*Anas* and *Aythya*, Anatidae) and a proposed evolutionary explanation. *Journal of Animal Ecology* 53: 75–92.

Townsend, G. H. 1966. A study of waterfowl nesting on the Saskatchewan River delta. *Canadian Field-Naturalist* 80: 74–88.

Williams, C. S., and W. H. Marshall. 1938. Duck nesting studies, Bear River Migratory Bird Refuge, Utah, 1937. *Journal of Wildlife Management* 2: 29–48.

Yocom, C. F. 1951. *Waterfowl and Their Food Plants in Washington.* University of Washington Press, Seattle.

Single- and Dual-Species Studies

Muscovy Duck

Blake, E. R. *Manual of Neotropical Birds.* Vol. 1. University of Chicago Press, Chicago (muscovy duck, p. 254).

Bolen, E. G. 1983. *Cairina moschata* (pato real, aliblanco, muscovy duck). Pp. 554–556 in D. H. Janzen, ed. *Costa Rican Natural History.* University of Chicago Press, Chicago IL.

Brush, T. 2005. *Nesting Birds of a Tropical Frontier: The Lower Rio Grande Valley of Texas.* Texas A&M University Press, College Station.

Brush, T. 2009. Breeding expansions and new breeding records of birds in Tamaulipas, Mexico. *Southwestern Naturalist* 54: 91–96.

Brush, T., and J. C. Eitniear. 2002. Status and recent nesting of muscovy duck (*Cairina moschata*) in the Rio Grande Valley, Texas. *Bulletin of the Texas Ornithological Society* 35: 12–14.

Cruz-Nieto, M. A. 1991. La situación actual del pato real (*Cairina moschata*) en México (proyecto cajones de anidamiento). *Memoria de III Symposium Internacional del Fauna Silvestre*, Universidad Autonóma de Tamaulipas, Victoria, Tamaulipas, México.

Del Toro, M. A. 1971. *Las Aves del Estado de Chiapas.* Instituto Historia Natural del Estado de Chiapas, Mexico. 270 pp.

Donkin, R. A. 1989. *The Muscovy Duck*, Cairina moschata domestica: *Origins, Dispersal, and Associated Aspects of the Geography of Domestication.* A. A. Balkema, Rotterdam, Netherlands.

Eitniear, J. C., A. Aragon-Tapia, and J. T. Baccus. 1998. Unusual nesting of muscovy duck *Cairina moschata* in northeastern Mexico. *Texas Journal of Science* 50: 173–175.

Fischer, D. H., J. Sanchez, M. McCoy, and E. G. Bolen. 1982. Aggressive displays of male muscovy ducks. *Brenesia* 19/20: 541–544.

Gómez-Dallmeier, F., and A. T. Cringan. 1989. *Biology, Conservation, and Management of Waterfowl in Venezuela.* Editorial ex Libris, Caracas, Venezuela.

Haverschmidt, F. 1968. *Birds of Surinam.* Oliver and Boyd, Edinburgh, UK.

Hill, S. L, and W. L. Brown. 1986. *A Guide to the Birds of Colombia.* Princeton University Press, Princeton, NJ (muscovy duck, pp. 85–86).

Hoffmann, E. 1992. A natural history of the South American pato *Cairina moschata:* The muscovy duck. *Ninth International Symposium on Waterfowl,* Pisa, Italy.

Hoffmann, E. 2005. Muscovy duck. Pp. 453–455 in J. Kear, ed. *Ducks, Geese, and Swans,* Vol. 2. Oxford University Press, Oxford, UK.

Howell, S. N. G., and S. Webb. 1995. *A Guide to the Birds of Mexico and Northern Central America.* Oxford University Press, Oxford, UK (muscovy duck, pp. 157–158).

Leopold, S. 1959. *Wildlife of Mexico: The Game Birds and Mammals.* University of California Press, Berkeley.

Markum, D. E., and G. A. Baldassarre. 1989. Breeding biology of muscovy ducks using nest boxes in Mexico. *Wilson Bulletin* 101: 621–626.

Munroe, B. L. J. 1968. *A Distributional Survey of the Birds of Honduras.* Ornithological Monographs, American Ornithologists' Union. No. 7. 758 pp.

Pranty, B., K. A. Radamaker, and G. Kennedy. 2006. *Birds of Florida.* Lone Pine Publishing International, Auburn, WA.

Raud, H., and J. M. Faure. 1988. Descriptive study of sexual behavior of male muscovy ducks. *Biology of Behavior* 13: 175.

Ridgely, R. S., and J. A Gwyne. 1989. *A Guide to the Birds of Panama.* Princeton University Press, Princeton, NJ (muscovy duck, p. 78).

Ridgely, R. S., and P. J. Greenfield. 2001. *The Birds of Ecuador.* Cornell University Press, Ithaca, NY (muscovy duck, pp. 53–54).

Rojas, P. M. 1954. Los patos silvestres en Mexico (su identificacion distribucion y notas relativasa su biologia). *Revista de la Sociedad Mexicana de Historia Natural* 15 (1–4): 95–107.

Sibley, C. L. 1967. The wild muscovy duck. *Modern Game Breeding* 3: 16–17.

Sick, H. 1993. *Birds in Brazil: A Natural History.* Princeton University Press, Princeton, NJ (muscovy duck, pp. 182–183).

Stai, S. M. 1999. Preliminary observations on the mating system in wild muscovy ducks (*Cairina moschata*). *Proceedings of the Neotropical Waterfowl Symposium: VI Neotropical Ornithological Congress,* Monterrey, Nuevo León, México.

Stiles, F. G., and A. F. Skutch. 1989. *A Guide to the Birds of Costa Rica.* Cornell University Press, Ithaca, NY (muscovy duck, p. 92).

Todd, F. S. 1996. *Natural History of the Waterfowl.* Ibis Publishing Co., Vista, CA (muscovy duck, pp. 254–255).

Wetmore, A. 1965. *The Birds of the Republic of Panama. Pt. 1, Tinamidae (Tinamous) to Rhynchopidae (Skimmers).* Smithsonian Miscellaneous Collections 150. 483 pp. (muscovy duck, pp. 134–137).

Whitley, G. R. 1973. The muscovy duck in Mexico. *Anthropological Journal of Canada* 11: 2–8.

Woodward, E. R. 1982. Some aspects of the ecology of the muscovy duck in Mexico. MS thesis, Texas Tech University, Lubbock.

Woodyard, E. R., and E. G. Bolen. 1984. Ecological studies of muscovy ducks in Mexico. *Southwestern Naturalist* 29: 453–461.

Wood Duck

Armbruster, J. S. 1982. Wood duck displays and pairing chronology. *Auk* 99: 116–122.

Baker, J. L. 1970. Wood duck brood survival on the Noxubee National Wildlife Refuge. *Proceedings of the Southeastern Association of Game and Fish Commissioners* 24: 104–108.

Baker, J. L. 1971. Wood duck (*Aix sponsa*) production from nest boxes and brood studies on the Noxubee National Wildlife Refuge. PhD dissertation, Mississippi State University, Starkville.

Ball, I. J., D. S. Gilmer, L. M. Cowardin, and J. H. Riechmann. 1975. Survival of wood duck and mallard broods in north-central Minnesota. *Journal of Wildlife Management* 39: 776–780.

Barras, S. C., R. M. Kaminski, and L. A. Brennan. 1996. Acorn selection by female wood ducks. *Journal of Wildlife Management* 60: 592–602.

Bartonek, J. C., J. T. Beall, and J. E. Cornely. 1990. Distribution, status, and harvest of wood ducks in the Pacific Flyway. Pp. 127–134 in L. H. Fredrickson, G. V. Burger, S. P. Havera, D. A. Graber, R. E. Kirby, and T. S. Taylor, eds. *Proceedings of the 1988 North American Wood Duck Symposium, St. Louis, Missouri.* Published by the symposium, St. Louis, MO.

Bellrose, F. C. 1976. The comeback of the wood duck. *Wildlife Society Bulletin* 4: 107–110.

Bellrose, F. C. 1990. The history of wood duck management. Pp. 13–20 in L. H. Fredrickson et al., eds. *Proceedings of the 1988 North American Wood Duck Symposium, St. Louis, Missouri.* Published by the symposium, St. Louis, MO.

Bellrose, F. C., and D. J. Holm. 1994. *Ecology and Management of the Wood Duck*. Stackpole Books, Mechanicsburg, PA. (Includes about 750 references.)

Bellrose, F. C., K. L. Johnson, and T. U. Meyers. 1964. Relative value of natural cavities and nesting houses for wood ducks. *Journal of Wildlife Management* 28: 661–676.

Bonar, R. L. 2000. Availability of pileated woodpecker cavities and use by other species. *Journal of Wildlife Management* 64: 52–59.

Bowers, E. F., and R. B. Hamilton. 1977. Derivation of northern wood ducks harvested in southern states of the Mississippi Flyway. *Proceedings of the Southeastern Association of Fish and Wildlife Agencies* 31: 90–98.

Bowers, E. F., and F. W. Martin. 1975. Managing wood ducks by population units. *Transactions of the North American Wildlife and Natural Resources Conference* 40: 300–324.

Breckenridge, W. J. 1956. Nesting study of wood ducks. *Journal of Wildlife Management* 20: 16–21.

Briggs, R. L. 1978. Wood ducks gathering acorns. *North American Bird Bander* 3: 102.

Brisbin, I. L., Jr., G. C. White, P. B. Bush, and L. A. Mayack. 1986. Sigmoid growth analyses of wood ducks: The effects of sex, dietary protein, and cadmium on parameters of the Richards model. *Growth* 50: 41–50.

Brisbin, I. L., Jr., G. C. White, P. B. Bush, and L. A. Mayack. 1987. Growth characteristics of wood ducks from two southeastern breeding locations. *Wilson Bulletin* 99: 91–94.

Conner, R. N., C. E. Shackelford, D. Saenz, and R. R. Schaefer. 2001. Interactions between pileated woodpeckers and wood ducks. *Wilson Bulletin* 113: 250–253.

Cottrell, S. D., H. H. Prince, and P. I. Padding. 1990. Nest success, duckling survival, and brood habitat selection of wood ducks in a Tennessee riverine system. Pp. 191–197 in L. H. Fredrickson, et al., eds. *Proceedings of the 1988 North American Wood Duck Symposium, St. Louis, Missouri*. Published by the symposium, St. Louis, MO.

Cringan, A. T. 1971. Status of the wood duck in Ontario. *Transactions of the North American Wildlife and Natural Resources Conference* 36: 296–311.

Cunningham, E. R. 1969. A three-year study of the wood duck on the Yazoo National Wildlife Refuge. *Proceedings Annual Conference of the Southeastern Association of Game and Fish Commissioners* 22: 145–155.

Davis, J. B., B. D. Leopold, R. M. Kaminski, and R. R. Cox, Jr. 2009. Wood duck duckling mortality and habitat implications in floodplain systems. *Wetlands* 29: 607–614.

Davis, J. B., R. R. Cox, Jr., R. M. Kaminski, and B. D. Leopold. 2007. Survival of wood duck ducklings and broods in Mississippi and Alabama. *Journal of Wildlife Management* 71: 507–517.

Delnicki, D. [E]., and K. J. Reinecke. 1986. Mid-winter food use and body weights of mallards and wood ducks in Mississippi. *Journal of Wildlife Management* 50: 43–51.

Dennis, D. G. 1990. Status of wood ducks in Canada. Pp. 135–137 in L. H. Fredrickson, et al., eds. *Proceedings of the 1988 North American Wood Duck Symposium, St. Louis, Missouri*. Published by the symposium, St. Louis, MO.

Dilger, W. C., and P. A. Johnsgard. 1959. Comments on "species recognition" with special reference to the wood duck and the mandarin duck. *Wilson Bulletin*, 71: 46–53.

Doty, H. A., and A. D. Kruse. 1972. Techniques for establishing local breeding populations of wood ducks. *Journal of Wildlife Management* 36: 428–435.

Drobney, R. D. 1980. Reproductive bioenergetics of wood ducks. *Auk* 97: 480–490.

Drobney, R. D., and L. H. Fredrickson. 1979. Food selection by wood ducks in relation to breeding status. *Journal of Wildlife Management* 43: 109–120.

Fendley, T. T. 1980. Incubating wood duck and hooded merganser hens killed by black rat snakes. *Wilson Bulletin* 92: 526–527.

Folk, T. H., and G. R. Hepp. 2003. Effects of habitat use and movement patterns on incubation behavior of female wood ducks (*Aix sponsa*) in southeast Alabama. *Auk* 120: 1159–1167.

Fredrickson, L. H., and J. L. Hansen. 1983. Second broods in wood ducks. *Journal of Wildlife Management* 47: 320–326.

Fredrickson, L. H., G. V. Burger, S. P. Havera, D. A. Graber, R. E. Kirby, and T. S. Taylor, eds. *Proceedings of the 1988 North American Wood Duck Symposium, St. Louis, Missouri*. Published by the symposium, St. Louis, MO.

Gilmer, D. S., I. J. Ball, L. M. Cowardin, J. E. Mathisen, and J. H. Reichmann. 1978. Natural cavities used by wood ducks in north-central Minnesota. *Journal of Wildlife Management* 42: 288–298.

Gilmer, D. S., R. E. Kirby, I. J. Ball, and J. H. Reichmann. 1977. Post-breeding activities of mallards and wood ducks in north-central Minnesota. *Journal of Wildlife Management* 41: 345–359.

Grice, D., and J. P. Rogers. 1965. *The Wood Duck in Massachusetts*. Federal Aid in Wildlife Restoration Project W-19-R. Massachusetts Division of Fisheries and Game, Westborough.

Hall, M. S. 1969. Unusual site for a wood duck nest. *New York Fish and Game Journal* 16: 217.

Haramis, G. M. 1990. Breeding ecology of the wood duck: A review. Pp. 45–60 in L. H. Fredrickson, et al., eds. *Proceedings of the 1988 North American Wood Duck Symposium, St. Louis, Missouri*. Published by the symposium, St. Louis, MO.

Haramis, G. M., and D. Q. Thompson. 1985. Density-production characteristics of box-nesting wood ducks in a northern green-tree impoundment. *Journal of Wildlife Management* 49: 429–436.

Hartke, K. M., and G. R. Hepp. 2004. Habitat use and preferences of breeding female wood ducks. *Journal of Wildlife Management* 68: 84–93.

Hein, D., and A. O. Haugen. 1966a. Autumn roosting flight counts as an index to wood duck abundance. *Journal of Wildlife Management* 30: 657–668.

Hein, D., and A. O. Haugen. 1966b. Illumination and wood duck roosting flights. *Wilson Bulletin* 78: 301–308.

Heitmeyer, M. E., and L. H. Fredrickson. 1990. Abundance and habitat use of wood ducks in the Mingo Swamp of southeastern Missouri. Pp. 141–151 in L. H. Fredrickson, et al., eds. *Proceedings of the 1988 North American Wood Duck Symposium, St. Louis, Missouri*. Published by the symposium, St. Louis, MO.

Henny, C. J. 1970. *Winter Bandings of Mallards, Black Ducks, Wood Ducks, Pintails, Canvasbacks, Redheads, and Scaup in 1967, 1968, and 1969*. Administrative Report 197. U.S. Bureau of Sport Fisheries and Wildlife, Laurel, MD.

Hepp, G. R., and J. D. Hair. 1977. Wood duck brood mobility and utilization of beaver pond habitats. *Proceedings of the Southeastern Association of Fish and Wildlife Agencies* 31: 216–225.

Hepp, G. R., and J. E. Hines. 1991. Factors affecting winter distribution and migration distance of wood ducks from southern breeding populations. *Condor* 93: 884–891.

Hepp, G. R., and R. A. Kennamer. 1992. Characteristics and consequences of nest-site fidelity in wood ducks. *Auk* 109: 812–818.

Hepp, G. R., and R. A. Kennamer. 1993. Effects of age and experience on reproductive performance of wood ducks. *Ecology* 74: 2027–2036.

Hepp, G. R., and R. A. Kennamer. 2011. Date of nest initiation mediates incubation costs of wood ducks (*Aix sponsa*). *Auk* 128: 258–264.

Hepp, G. R., D. J. Stangohr, L. A. Baker, and R. A. Kennamer. 1987. Factors affecting variation in the egg and duckling components of wood ducks. *Auk* 104: 435–443.

Hepp, G. R., R. A. Kennamer, and M. H. Johnson. 2006. Maternal effects in wood ducks: incubation temperature influences incubation period and neonate phenotype. *Functional Ecology* 20: 307–314.

Hepp, G. R., R. A. Kennamer, and W. F. Harvey, IV. 1989. Recruitment and natal philopatry of wood ducks. *Ecology* 70: 897–903.

Hepp, G. R., R. A. Kennamer, and W. F. Harvey, IV. 1990. Incubation as a reproductive cost in female wood ducks. *Auk* 107: 756–764.

Hepp, G. R., R. T. Hoppe, and R. A. Kennamer. 1987a. Population parameters and philopatry of breeding female wood ducks. *Journal of Wildlife Management* 51: 401–404.

Hester, F. E., and J. Dermid. 1973. *The World of the Wood Duck*. J. B. Lippincott, Philadelphia, PA.

Heusmann, H. W. 1975. Several aspects of the nesting biology of yearling wood ducks. *Journal of Wildlife Management* 39: 503–507.

Heusmann, H. W. 2000. Production from wood duck nest boxes as a proportion of the harvest in Massachusetts. *Wildlife Society Bulletin* 28: 1046–1049.

Hipes, D. L., and G. R. Hepp. 1995. Nutrient-reserve dynamics of breeding male wood ducks. *Condor* 97: 451–460.

Hocutt, G. E., and R. W. Dimmick. 1971. Summer food habits of juvenile wood ducks in east Tennessee. *Journal of Wildlife Management* 35: 286–292.

Johnsgard, P. A. 1968. Some putative mandarin duck hybrids. *Bulletin British Ornithologist's Club* 88: 140–148. http://digitalcommons.unl.edu/johnsgard/26/

Johnson, N. F. 1971. Effects of dietary protein on wood duck growth. *Journal of Wildlife Management* 35: 798–802.

Kaminski, R. M., J. B. Davis, H. W. Essig, P. D. Gerard, and K. J. Reinecke. 2003. True metabolizable energy for wood ducks from acorns compared to other waterfowl foods. *Journal of Wildlife Management* 67: 542–550.

Kaminski, R. M., R. W. Alexander, and B. D. Leopold. 1993. Wood duck and mallard microhabitats in Mississippi hardwood bottomlands. *Journal of Wildlife Management* 57: 562–570.

Kennamer, R. A., and G. R. Hepp. 1987. Frequency and timing of second broods in wood ducks. *Wilson Bulletin* 99: 655–662.

Kennamer, R. A., W. F. Harvey, IV, and G. R. Hepp. 1990. Embryonic development and nest attentiveness of wood ducks during egg laying. *Condor* 92: 587–592.

Kirby, R. E., and L. H. Fredrickson. 1990. Molts and plumages of the wood duck. Pp. 29–33 in L. H. Fredrickson, et al., eds. *Proceedings of the 1988 North American Wood Duck Symposium, St. Louis, Missouri.* Published by the symposium, St. Louis, MO.

Korschgen, C. E., and L. H. Fredrickson. 1976. Comparative displays of yearling and adult male wood ducks. *Auk* 93: 793–807.

Ladd, W. N., Jr. 1990. Status of wood ducks in the Central Flyway. Pp. 121–126 in L. H. Fredrickson, et al., eds. *Proceedings of the 1988 North American Wood Duck Symposium, St. Louis, Missouri.* Published by the symposium, St. Louis, MO.

Landers, J. L., T. T. Fendley, and A. S. Johnson. 1977. Feeding ecology of wood ducks in South Carolina. *Journal of Wildlife Management* 41: 118–127.

Leopold, F. 1951. A study of nesting wood ducks in Iowa. *Condor* 53: 209–220.

Lowney, M. S., and E. P. Hill. 1989. Wood duck nest sites in bottomland hardwood forests of Mississippi. *Journal of Wildlife Management* 53: 378–382.

Mason, P. J., and J. L. Dusi. 1983. A ground-nesting wood duck. *Auk* 100: 506.

McGilvrey, F. B. 1968. *A Guide to Wood Duck Production Habitat Requirements.* Resource Publication 60. US Bureau of Sport Fisheries and Wildlife, Washington, DC.

McGilvrey, F. B. 1969. Survival of wood duck broods. *Journal of Wildlife Management* 33: 73–76.

McIlquham, C. J., and B. R. Bacon. 1989. Wood duck nest on a muskrat house. *Journal of Field Ornithology* 60: 84–85.

Moorman, T. E., and G. A. Baldassarre. 1988. Incidence of second broods by wood ducks in Alabama and Georgia. *Journal of Wildlife Management* 52: 426–431.

Naylor, A. E. 1960. The wood duck in California with special reference to the use of nest boxes. *California Fish and Game* 46: 241–269.

Nichols, J. D., and F. A. Johnson. 1990. Wood duck population dynamics: A review. Pp. 83–105 in L. H. Fredrickson, et al., eds. *Proceedings of the 1988 North American Wood Duck Symposium, St. Louis, Missouri.* Published by the symposium, St. Louis, MO.

Nielsen, C. L. R., R. J. Gates, and E. H. Zwicker. 2007. Projected availability of natural cavities for wood ducks in southern Illinois. *Journal of Wildlife Management* 71: 875–883.

Parr, D. E., and M. D. Scott. 1978. Analysis of roosting counts as an index to wood duck population size. *Wilson Bulletin* 90: 423–437.

Parr, D. E., M. D. Scott, and D. D. Kennedy. 1979. Autumn movements and habitat use by wood ducks in southern Illinois. *Journal of Wildlife Management* 43: 102–108.

Perry, H. R., Jr. 1977. Wood duck roost utilization of northeastern North Carolina swamps. *Proceedings of the Southeastern Association of Fish and Wildlife Agencies* 31: 307–311.

Ransom, D., Jr., R. L. Honeycutt, and R. D. Slack. 2001. Population genetics of southeastern wood ducks. *Journal of Wildlife Management* 65: 745–754.

Ridlehuber, K. T., B. W. Cain, and N. J. Silvy. 1990. Movements, habitat use, and survival of wood duck broods in east-central Texas. *Proceedings of the Southeastern Association of Fish and Wildlife Agencies* 44: 284–294.

Robb, J. R., and T. A. Bookhout. 1995. Factors influencing wood duck use of natural cavities. *Journal of Wildlife Management* 59: 372–383.

Rogers, J. P., and J. L. Hansen. 1967. Second broods in the wood duck. *Bird Banding* 38: 234–235.

Rohwer, F. C., and H. W. Heusmann. 1991. Effects of brood size and age on survival of female wood ducks. *Condor* 93: 817–824.

Rollin, N. 1957. Incubation by drake wood duck in eclipse plumage. *Condor* 59: 263–265.

Ryan, D. C., R. J. Kawula, and R. J. Gates. 1998. Breeding biology of wood ducks using natural cavities in southern Illinois. *Journal of Wildlife Management* 62: 112–123.

Shurtleff, L. I., and C. Savage. 1996. *The Wood Duck and the Mandarin: The Northern Wood Ducks*. University of California Press, Berkeley.

Soulliere, G. J. 1990. Review of wood duck nest cavity characteristics. Pp. 153–162 in L. H. Fredrickson, et al., eds. *Proceedings of the 1988 North American Wood Duck Symposium, St. Louis, Missouri*. Published by the symposium, St. Louis, MO.

Stewart, P. A. 1958. Local movements of wood ducks (*Aix sponsa*). *Auk* 75: 157–168.

Tabberer, D. K., J. D. Newsom, P. E. Schilling, and H. A. Bateman. 1971. The wood duck roost count as an index to wood duck abundance in Louisiana. *Proceedings of the Southeastern Association of Game and Fish Commissioners* 25: 254–261.

Thompson, J. D., and G. A. Baldassarre. 1988. Postbreeding habitat preference of wood ducks in northern Alabama. *Journal of Wildlife Management* 52: 80–85.

Thompson, J. D., and G. A. Baldassarre. 1989. Postbreeding dispersal by wood ducks in northern Alabama with reference to early hunting seasons. *Wildlife Society Bulletin* 17: 142–146.

Thompson, S. C., and S. B. Simmons. 1990. Characteristics of second clutches in California wood ducks. Pp. 171–177 in L. H. Fredrickson, et al., eds. *Proceedings of the 1988 North American Wood Duck Symposium, St. Louis, Missouri*. Published by the symposium, St. Louis, MO.

Trefethen, J. B., ed. 1966. *Wood Duck Management and Research: A Symposium. Theme: Emphasizing Management of Forests for Wood Ducks*. Wildlife Management Institute, Washington, DC.

Weier, R. W. S. 1966. Survey of wood duck nest sites on Mingo Wildlife Refuge in southeast Missouri. Pp. 91–108 in J. B. Trefethen, ed. *Wood Duck Management and Research: A Symposium. Theme: Emphasizing Management of Forests for Wood Ducks*. Washington, DC: Wildlife Management Institute, Washington, DC.

Weller, M. W., ed. 1988. *Waterfowl in Winter*. University of Minnesota Press, Minneapolis.

Yetter, A. P., S. P. Havera, and C. S. Hine. 1999. Natural-cavity use by nesting wood ducks in Illinois. *Journal of Wildlife Management* 63: 630–638.

Zipko, S. J., and J. Kennington. 1977. A ground-nesting wood duck. *Auk* 94: 159.

Wood Ducks and Brood Parasitism

Bolen, E. G., and B. W. Cain. 1968. Mixed wood duck–tree duck clutch in Texas. *Condor* 70: 389–390.

Bouvier, J. M. 1974. Breeding biology of the hooded merganser in southwestern Québec, including interactions with common goldeneyes and wood ducks. *Canadian Field-Naturalist* 88: 323–330.

Clawson, R. L., G. W. Hartman, and L. H. Fredrickson. 1979. Dump nesting in a Missouri wood duck population. *Journal of Wildlife Management* 43: 347–355.

Doty, H. A., F. B. Lee, A. D. Kruse, J. W. Matthews, J. R. Foster, and P. M. Arnold. 1984b. Wood duck and hooded merganser nesting on Arrowwood NWR, North Dakota. *Journal of Wildlife Management* 48: 577–580.

Hansen, J. L. 1971. The role of nest boxes in the management of the wood duck on Mingo National Wildlife Refuge. MS thesis, University of Missouri, Columbia.

Heusmann, H. W. 1972. Survival of wood duck broods from dump nests. *Journal of Wildlife Management* 36: 620–624.

Labuda, S. E. 1969. Tree duck–wood duck egg parasitism. *Bulletin of the Texas Ornithological Society* 3: 28.

Mallory, M. L., A. Taverner, B. Bower, and D. Crook. 2002. Wood duck and hooded merganser breeding success in nest boxes in Ontario. *Wildlife Society Bulletin* 30: 310–316.

Morse, T. E., and H. M. Wight. 1969. Dump nesting and its effect on production in wood ducks. *Journal of Wildlife Management* 33: 284–293.

Prince, H. H. 1965. The breeding ecology of wood duck (*Aix sponsa* L.) and common goldeneye (*Bucephala clangula* L.) in central New Brunswick. MS thesis, University of New Brunswick, Fredericton.

Prince, H. H. 1968. Nest sites used by wood ducks and common goldeneyes in New Brunswick. *Journal of Wildlife Management* 32: 489–500.

Robinson, S. S. 1983. Hooded merganser nesting in wood duck boxes in Wake County, N.C. *Chat* 47: 19–22.

Semel, B., and P. W. Sherman. 1992. Use of clutch size to infer brood parasitism in wood ducks. *Journal of Wildlife Management* 56: 495–499.

Semel, B., and P. W. Sherman. 2001. Intraspecific parasitism and nest-site competition in wood ducks. *Animal Behaviour* 61: 787–803.

Semel, B., P. W. Sherman, and S. M. Byers. 1988. Effects of brood parasitism and nest-box placement on wood duck breeding ecology. *Condor* 90: 920–930.

Zicus, M. C. 1990b. Nesting biology of hooded mergansers using nest boxes. *Journal of Wildlife Management* 54: 637–643.

Eurasian Wigeon

Edgell, M. C. R. 1984. Trans-hemispheric movements of Holarctic Anatidae: The Eurasian wigeon (*Anas penelope* L.) in North America. *Journal of Biogeography* 11: 27–39.

Fournier, M. A., and J. E. Hines. 1996. Second record and possible breeding of the Eurasian wigeon, *Anas penelope*, in the District of Mackenzie, Northwest Territories. *Canadian Field-Naturalist* 110: 336–337.

Hasbrouck, E. M. 1944. Apparent status of the European wigeon in North America. *Auk* 61: 93–104.

Merrifield, K. 1993. Eurasian × American wigeons in western Oregon. *Western Birds* 24: 105–107.

Mitchell, C. 2005. Eurasian wigeon. Pp. 499–502 in J. Kear, ed. *Ducks, Geese, and Swans.* Vol. 2. Oxford University Press, Oxford, UK.

Ogilvie, M. A. 1975. *Ducks of Britain and Europe.* T. & A. D. Poyser, Berkhamsted, UK.

Owen, M. 1973. The winter feeding ecology of wigeon at Bridgewater Bay, Somerset. *Ibis* 115: 227–243.

Owen, M. 1977. *Wildfowl of Europe.* Macmillan, London, UK (Eurasian wigeon, pp. 156–162).

Roselaar, C. S. 1977. Wigeon. Pp. 473–483 in S. Cramp and K. E. L. Simmons. *Handbook of the Birds of Europe, the Middle East, and North Africa: The Birds of the Western Palearctic, Volume 1, Ostrich to Ducks.* Oxford University Press, Oxford, UK (*A. penelope*).

American Wigeon

Arnold, T. W., and R. G. Clark. 1996. Survival and philopatry of female dabbling ducks in southcentral Saskatchewan. *Journal of Wildlife Management* 60: 560–568.

Bartonek, J. C. 1972. Summer foods of American wigeon, mallards, and a green-winged teal near Great Slave Lake, N.W.T. *Canadian Field-Naturalist* 86: 373–376.

Eamer, J. 1981. Winter feeding ecology of mallards and American wigeon along the east coast of Vancouver Island, BC. *Pacific Seabird Group Bulletin* 8: 88.

Esler, D., and J. B. Grand. 1994. Comparison of age determination techniques for female northern pintails and American wigeon in spring. *Wildlife Society Bulletin* 22: 260–264.

Fuller, R. W., and N. E. King. 1964. American wigeon and shoveler breeding in Vermont. *Auk* 81: 86–87.

Goudie, R. I. 1985. Range extension of the American wigeon, *Anas americana*, to the island of Newfoundland. *Canadian Field-Naturalist* 99: 533.

Graves, K. L.1996. Feeding symbiosis between the American coot and American wigeon. *Oregon Birds* 21: 9.

Hitchcock, R. R., R. Balcomb, and R. J. Kendall. 1993. Migration chronology of American wigeon in Washington, Oregon, and California. *Journal of Field Ornithology* 64: 96–101.

Kendall, R. J., L. W. Brewer, R. R. Hitchcock, and J. R. Mayer. 1992. American wigeon mortality associated with turf application of Diazinon AG500. *Journal of Wildlife Diseases* 28: 263–267.

Knapton, R. W. 1992. The American wigeon at Long Point: A species on the increase? *Long Point Observatory Newsletter* 24: 15.

Knapton, R. W., and B. Knudsen. 1978. Food piracy by American wigeons on American coots. *Canadian Field-Naturalist* 92: 403–404.

McClanahan, R. C. 1942. Male baldpate attending young. *Auk* 59: 589.

Merrifield, K. 1993. Eurasian × American wigeons in western Oregon. *Western Birds* 24: 105–107 (three wild hybrids).

Rienecker, W. C. 1976. Distribution, harvest, and survival of American wigeon banded in California. *California Fish and Game* 62: 141–153.

Ryan, M. R. 1981. Evasive behavior of American coots to kleptoparasitism by waterfowl. *Wilson Bulletin* 93: 274–275.

Soutiere, E. C., H. S. Myrick, and E. G. Bolen. 1972. Chronology and behavior of American widgeon wintering in Texas. *Journal of Wildlife Management* 36: 752–758.

Spencer, H. E., Jr. 1977. American wigeon breeding in Maine. *Auk* 94: 700.

Turnbull, R. E., and G. A. Baldassarre. 1987. Activity budgets of mallards and American wigeon wintering in east-central Alabama. *Wilson Bulletin* 99: 457–464.

Wishart, R. A. 1979. Indices of structural size and condition of American wigeon (*Anas americana*). *Canadian Journal of Zoology* 57: 2369–2374.

Wishart, R. A. 1983a. The behavioral ecology of the American wigeon (*Anas americana*) over its annual cycle. PhD dissertation, University of Manitoba, Winnipeg.

Wishart, R. A. 1983b. Pairing chronology and mate selection in the American wigeon (*Anas americana*). *Canadian Journal of Zoology* 61: 1733–1743.

Wishart, R. A. 1985. Moult chronology of American wigeon, *Anas americana*, in relation to reproduction. *Canadian Field-Naturalist* 99: 172–178.

Falcated Duck

Brewer, G. L. 2005. Falcated duck. Pp. 495–496 in J. Kear, ed. *Ducks, Geese, and Swans*. Volume 2. Oxford University Press, Oxford, UK.

Cheng, Tso-hsin. 1963. *China's Economic Fauna: Birds*. Peiping (Beijing), China: Science Publishing Society (translated by Joint Publication Research Service, Washington, DC, 1964).

Cheng, Tso-hsin. 1987. *A Synopsis of the Avifauna of China*. Science Press and Paul Perry, Berlin, Germany (falcated duck, pp. 55–56).

De Schauensee, R. M. 1984. *The Birds of China*. Smithsonian Institution Press, Washington, DC (falcated duck, p. 144).

Roselaar, C. S. 1977. Wigeon. Pp. 483–485 in S. Cramp and K. E. L. Simmons. *Handbook of the Birds of Europe, the Middle East, and North Africa: The Birds of the Western Palearctic, Volume 1, Ostrich to Ducks*. Oxford University Press, Oxford, UK.

Gadwall

Ankney, C. D., and R. T. Alisauskas. 1991. Nutrient-reserve dynamics and diet of breeding female gadwalls. *Condor* 93: 799–810.

Blohm, R. J. 1979. The breeding ecology of the gadwall in southern Manitoba. PhD dissertation, University of Wisconsin–Madison, Madison.

Blohm, R. J. 1982. Differential occurrence of yearling and adult male gadwalls in pair bonds. *Auk* 99: 378–379.

Borden, R., and H. A. Hochbaum. 1966. Gadwall seeding in New England. *North American Wildlife and Natural Resources Conference Transactions* 31: 79–88.

Canning, D. J., and S. G. Herman. 1983. Gadwall breeding range expansion into western Washington. *Murrelet* 64: 27–31.

Chabreck, R. H. 1966. Molting gadwall (*Anas strepera*) in Louisiana. *Auk* 83: 664.

Duebbert, H. F. 1966. Island nesting of the gadwall in North Dakota. *Wilson Bulletin* 78: 12–25.

Duebbert, H. F., J. T. Lokemoen, and D. E. Sharp. 1983. Concentrated nesting of mallards and gadwalls on Miller Lake Island, North Dakota. *Journal of Wildlife Management* 47: 729–740.

Dwyer, T. J. 1974. Social behavior of breeding gadwall in North Dakota. *Auk* 91: 375–386.

Dwyer, T. J. 1975. Time budget of breeding gadwalls. *Wilson Bulletin* 87: 335–343.

Fox, T. 2005. Gadwall. Pp. 491–494 in J. Kear, ed. *Ducks, Geese, and Swans.* Vol. 2. Oxford University Press, Oxford, UK.

Gaston, G. R. 1991. Effects of environment and hunting on body condition of nonbreeding gadwalls (*Anas strepera*) in southeastern Louisiana. *Southwestern Naturalist* 36: 318–322.

Gaston, G. R., D. Walther, D. Shaheen, and J. D. Felley. 1989. Winter condition of gadwalls (*Anas strepera*) in southwestern Louisiana. *Proceedings of the Louisiana Academy of Science* 52: 55–61.

Gates, J. M. 1957. Autumn food habits of the gadwall in northern Utah. *Proceedings of the Utah Academy of Sciences, Arts, and Letters* 34: 69–71.

Gates, J. M. 1962. Breeding biology of the gadwall in northern Utah. *Wilson Bulletin* 74: 43–67.

Griffith, R. E. 1946. Nesting of gadwall and shoveler on the middle Atlantic coast. *Auk* 63: 436–438.

Henny, C. J., and N. E. Holgersen. 1974. Range expansion and population increase of the gadwall in eastern North America. *Wildfowl* 25: 95–101.

Hines, J. E., and G. J. Mitchell. 1983a. Breeding ecology of the gadwall at Waterhen Marsh, Saskatchewan. *Canadian Journal of Zoology* 61: 1532–1539.

Hines, J. E., and G. J. Mitchell. 1983b. Gadwall nest-site selection and nesting success. *Journal of Wildlife Management* 47: 1063–1071.

Hines, J. E., and G. J. Mitchell. 1984. Parasitic laying in nests of gadwalls. *Canadian Journal of Zoology* 62: 627–630.

Keith, L. B. 1961. *A Study of Waterfowl Ecology on Small Impoundments in Southeastern Alberta.* Wildlife Monographs, No. 6. The Wildlife Society, Bethesda, MD.

LeShack, C. R., and G. R. Hepp. 1995. Kleptoparasitism of American coots by gadwalls and its relationship to social dominance and food abundance. *Auk* 112: 429–435.

McKnight, S. K., and G. R. Hepp. 1998. Diet selectivity of gadwalls wintering in Alabama. *Journal of Wildlife Management* 62: 1533–1543.

Oring, L. W. 1966. Breeding biology of the gadwall, *Anas strepera* Linnaeus. PhD dissertation, University of Oklahoma, Norman.

Oring, L. W. 1968. Growth, molts, and plumages of the gadwall. *Auk* 85: 355–380.

Oring, L. W. 1969. Summer biology of the gadwall at Delta, Manitoba. *Wilson Bulletin* 81: 44–54.

Paulus, S. L. 1982. Feeding ecology of gadwalls in Louisiana in winter. *Journal of Wildlife Management* 46: 71–79.

Paulus, S. L. 1983. Dominance relations, resource use, and pairing chronology of gadwalls in winter. *Auk* 100: 947–952.

Paulus, S. L. 1984a. Activity budgets of nonbreeding gadwalls in Louisiana. *Journal of Wildlife Management* 48: 371–380.

Paulus, S. L. 1984b. Molts and plumages of gadwalls in winter. *Auk* 101: 887–889.

Peters, J., and K. E. Omland. 2007. Population structure and mitochondrial polyphyly in North American gadwalls (*Anas strepera*). *Auk* 124: 444–462.

Peters, J., Y. N. Zhuravlev, I. Fefelov, E. M. Humphries, and K. E. Omland. 2008. Multilocus phylogeography of a Holarctic duck: Colonization of North America from Eurasia by gadwall (*Anas strepera*). *Evolution* 62: 1469–1483.

Peters, J. L., G. L. Brewer, and L. M. Bowe. 2003. Extrapair paternity and breeding synchrony in gadwalls (*Anas strepera*) in North Dakota. *Auk* 120: 883–888.

Pietz, P. J., G. L. Krapu, D. A. Brandt, and R. R. Cox, Jr. 2003. Factors affecting gadwall brood and duckling survival in prairie pothole landscapes. *Journal of Wildlife Management* 67: 564–575.

Pietz, P. J., G. L. Krapu, D. A. Buhl, and D. A. Brandt. 2000. Effects of water conditions on clutch size, egg volume, and hatching mass of mallards and gadwalls in the Prairie Pothole region. *Condor* 102: 936–940.

Sayler, R. D., and M. A. Willams. 1997. Brood ecology of mallards and gadwalls nesting on islands in large reservoirs. *Journal of Wildlife Management* 61: 808–815.

Sedwitz, W. 1958. Six years (1947–1952) nesting of gadwall (*Anas strepera*) on Jones Beach, Long Island, NY. *Proceedings Linnaean Society of New York*, No. 66–70: 61–70.

Serie, J. R., and G. A. Swanson. 1976. Feeding ecology of breeding gadwalls on saline wetlands. *Journal of Wildlife Management* 40: 69–81.

Sousa, P. J. 1985. *Habitat Suitability Index Models: Gadwall (Breeding). Biological Report 82 (10.100).* Western Energy and Land Use Team, Division of Biological Services, US Fish and Wildlife Service, Washington, DC.

Sugden, L. G. 1973. *Feeding Ecology of Pintail, Gadwall, American Widgeon, and Lesser Scaup Ducklings in Southern Alberta.* Report Series 24. Canadian Wildlife Service, Ottawa, ON.

Szymczak, M. R., and E. A. Rexstad. 1991. Harvest distribution and survival of a gadwall population. *Journal of Wildlife Management* 55: 592–600.

Baikal Teal

Cheng, Tso-hsin. 1987. *A Synopsis of the Avifauna of China.* Science Press and Paul Perry, Berlin, Germany (Baikal teal, pp. 55).

Hatter, J. 1960. Baikal teal in British Columbia. *Condor* 62: 480.

Jewett, S. G., W. P. Taylor, W. T. Shaw, and J. W. Aldrich. 1953. *Birds of Washington State.* University of Washington Press, Seattle.

Maher, W. J. 1960. Another record of the Baikal teal in northwestern Alaska. *Condor* 62: 138.

Moores, N. 2005. Baikal teal. Pp. 605–608 in J. Kear, ed. *Ducks, Geese, and Swans.* Vol. 2. Oxford University Press, Oxford, UK.

Ogilvie, M. A. 1975. *Ducks of Britain and Europe.* T. & A. D. Poyser, Berkhamsted, UK.

Owen, M. 1977. *Wildfowl of Europe.* Macmillan, London, UK (Baikal teal, pp. 164–165).

Green-winged Teal

Baldassarre, G. A., and E. G. Bolen. 1986. Body weight and aspects of pairing chronology of green-winged teal and northern pintails wintering on the southern High Plains of Texas. *Southwestern Naturalist* 31: 361–366.

Baldassarre, G. A., E. E. Quinlan, and E. G. Bolen. 1988. Mobility and site fidelity of green-winged teal wintering on the southern High Plains of Texas. Pp. 483–493 in M. W. Weller, ed. *Waterfowl in Winter.* University of Minnesota Press, Minneapolis.

Baldassarre, G. A., R. J. Whyte, and E. G. Bolen. 1986. Body weight and carcass composition of green-winged teal on the southern High Plains of Texas. *Journal of Wildlife Management* 50: 420–426.

Bennett, J. W., and E. G. Bolen. 1978. Stress response in wintering green-winged teal. *Journal of Wildlife Management* 42: 81–86.

Euliss, N. H., Jr., and S. W. Harris. 1987. Feeding ecology of northern pintails and green-winged teal wintering in California. *Journal of Wildlife Management* 51: 724–732.

Fox, T. 2005. Eurasian and American green-winged teal. Pp. 609–613 in J. Kear, ed. *Ducks, Geese, and Swans.* Vol. 2. Oxford University Press, Oxford, England, UK.

Guillemain, M., J. Bertout, T. K. Christensen, H. Pöysä, V. Väänänen, P. Triplet, V. Schricke, and A. D. Fox. 2010. How many juvenile teal *Anas crecca* reach the wintering grounds? Flyway-scale survival rate inferred from wing age-ratios. *Journal of Ornithology* 151: 51–60.

Johnson, W. P., and F. C. Rohwer. 1998. Pairing chronology and agonistic behaviors of wintering green-winged teal and mallards. *Wilson Bulletin* 110: 311–315.

Johnson, W. P., and F. C. Rohwer. 2000. Foraging behavior of green-winged teal and mallards on tidal mudflats in Louisiana. *Wetlands* 20: 184–188.

Laurie-Ahlberg, C. C., and F. McKinney. 1979. The nod-swim display of male green-winged teal *Animal Behaviour* 27: 165–172.

McKinney, F. 1965. The displays of the American green-winged teal. *Wilson Bulletin* 77: 112–121.

McKinney, F., and P. Stolen. 1982. Extra-pair-bond courtship and forced copulation among captive green-winged teal (*Anas crecca carolinensis*). *Animal Behaviour* 30: 461–474.

McKinney, F., and S. Derrickson. 1979. Aerial scratching, leeches, and nasal saddles in green-winged teal. *Wildfowl* 30: 151–153.

Moisan, G. 1967. *The Green-winged Teal: Its Distribution, Migration, and Population Dynamics.* Special Scientific Report—Wildlife 100. US Fish and Wildlife Service, Washington, DC.

Munro, J. A. 1949. Studies of waterfowl in British Columbia: Green-winged teal. *Canadian Journal of Research* 27: 149–178.

Nummi, P. 1993. Food-niche relationships of sympatric mallards and green-winged teal. *Canadian Journal of Zoology* 71: 49–55.

Paquette, G. A., and C. D. Ankney. 1996. Wetland selection by American green-winged teal breeding in British Columbia. *Condor* 98: 27–33.

Quinlan, E. E., and G. A. Baldassarre. 1984. Activity budgets of nonbreeding green-winged teal on playa lakes in Texas. *Journal of Wildlife Management* 48: 838–845.

Rave, D. P., and G. A. Baldassarre. 1989. Activity budget of green-winged teal wintering in coastal wetlands of Louisiana. *Journal of Wildlife Management* 53: 753–759.

Rave, D. P., and G. A. Baldassarre. 1991. Carcass mass and composition of green-winged teal wintering in Louisiana and Texas. *Journal of Wildlife Management* 55: 457–481.

Rollo, J. D., and E. G. Bolen. 1969. Ecological relationships of blue and green-winged teal on the High Plains of Texas in early fall. *Southwestern Naturalist* 14: 171–188.

Tamisier, A. 1976. Diurnal activities of green-winged teal and pintail wintering in Louisiana. *Wildfowl* 27: 19–32.

Northern Mallard

Anderson, D. R. 1975. *Population Biology of the Mallard, V: Temporal and Geographic Estimates of Survival, Recovery and Harvest Rates.* Resource Publication 125. US Fish and Wildlife Service, Washington, DC.

Anderson, D. R., and K. P. Burnham. 1976. *Population Ecology of the Mallard, VI: The Effect of Exploitation on Survival.* Resource Publication 128. US Fish and Wildlife Service, Washington, DC.

Anderson, D. R., and K. P. Burnham. 1978. Effect of restrictive and liberal hunting regulations on annual survival rates of the mallard in North America. *Transactions of the North American Wildlife and Natural Resources Conference* 43: 181–186.

Ball, I. J., D. S. Gilmer, L. M. Cowardin, and J. H. Riechmann. 1975. Survival of wood duck and mallard broods in north-central Minnesota. *Journal of Wildlife Management* 39: 776–780.

Bjarvall, A. 1968. The hatching and nest-exodus behaviour of mallard. *Wildfowl* 19: 70–80.

Bjarvall, A. 1969. Unusual cases of re-nesting mallards. *Wilson Bulletin* 81: 94–96

Batt, B. D. J., and H. H. Prince. 1978. Some reproductive parameters of mallards in relation to age, captivity, and geographic origin. *Journal of Wildlife Management* 42: 834–842.

Batt, B. D. J., and H. H. Prince. 1979. Laying dates, clutch size, and egg weight of captive mallards. *Condor* 81: 35–41.

Bellrose, F. C. 1972. Mallard migration corridors as revealed by population distribution, banding, and radar. Pp. 3–26 in *Population Ecology of Migratory Birds.* Wildlife Research Report 2. US Bureau of Sport Fisheries and Wildlife, Washington, DC.

Bergan, J. F., and L. M. Smith. 1993. Survival rates of female mallards wintering in the Playa Lakes Region. *Journal of Wildlife Management* 57: 570–577.

Bishop, R. A., and R. Barratt. 1970. Use of artificial nest baskets by mallards. *Journal of Wildlife Management* 34: 734–738.

Bishop, R. A., D. D. Humburg, and R. D. Andrews. 1978. Survival and homing of female mallards. *Journal of Wildlife Management* 42: 192–196.

Bjärvall, A. 1968. The hatching and nest-exodus behaviour of mallard. *Wildfowl* 19: 70–80.

Boos, J. D., T. D. Nudds, and K. Sjöberg. 1989. Posthatch brood amalgamation by mallards. *Wilson Bulletin* 101: 503–505.

Boyd, H. 1961. The flightless period of the mallard in England. *Wildfowl Trust Annual Report* 12: 140–143.

Brasher, M. G., T. W. Arnold, J. H. Devries, and R. M. Kaminski. 2006. Breeding-season survival of male and female mallards in Canada's prairie-parklands. *Journal of Wildlife Management* 70: 805–811.

Calverley, B. K., and D. A. Boag. 1977. Reproductive potential in parkland and arctic-nesting populations of mallards and pintails (Anatidae). *Canadian Journal of Zoology* 55: 1242–1251.

Cheng, K. M., J. T. Burns, and F. McKinney. 1980. Forced copulation in captive mallards, 1: Fertilization of eggs. *Auk* 97: 875–879.

Cheng, K. M., J. T. Burns, and F. McKinney. 1982. Forced copulation in captive mallards (*Anas platyrhynchos*), 2: Temporal factors. *Animal Behaviour* 30: 695–699.

Cheng, K. M., J. T. Burns, and F. McKinney. 1983. Forced copulation in captive mallards, 3: Sperm competition. *Auk* 100: 302–310.

Chouinard, M. P., and T. W. Arnold. 2007. Survival and habitat use of mallard (*Anas platyrhynchos*) broods in the San Joaquin Valley, California. *Auk* 124: 1305–1316.

Chura, N. J. 1961. Food availability and preferences of juvenile mallards. *Transactions of the North American Wildlife and Natural Resources Conference* 26: 121–133.

Corman, T. E. 2005. Mallard (*Anas platyrhynchos*). Pp. 58–59 in E. Corman and C. Wise-Gervais, eds., *Arizona Breeding Bird Atlas*. University of New Mexico Press, Albuquerque.

Coulter, M. W. 1953. Mallard nesting in Maine. *Auk* 70: 490.

Coulter, M. W. 1954. Some observations of mallards in central Maine. *Maine Audubon Society Bulletin* 10: 20–23.

Cowardin, L. M., and D. H. Johnson. 1979. Mathematics and mallard management. *Journal of Wildlife Management* 43: 18–35.

Cowardin, L. M., D. S. Gilmer, and C. W. Shaiffer. 1985. *Mallard Recruitment in the Agricultural Environment of North Dakota.* Wildlife Monographs 92. The Wildlife Society, Bethesda, MD.

Cox, R. R., Jr., M. A. Hanson, C. C. Roy, N. H. Euliss, Jr., D. H. Johnson, and M. G. Butler. 1998. Mallard duckling growth and survival in relation to aquatic invertebrates. *Journal of Wildlife Management* 62: 124–133.

Dabbert, C. B., and T. E. Martin. 2000. Diet of mallards wintering in greentree reservoirs in southeastern Arkansas. *Journal of Field Ornithology* 71: 423–428.

Davis, B. E., A. D. Afton, and R. R. Cox, Jr. 2011. Factors affecting winter survival of female mallards in the lower Mississippi Alluvial Valley. *Waterbirds* 34: 186–194.

Delnicki, D., and K. J. Reinecke. 1986. Mid-winter food use and body weights of mallards and wood ducks in Mississippi. *Journal of Wildlife Management* 50: 43–51.

Drewien, R. C., and L. F. Fredrickson. 1970. High density mallard nesting on a South Dakota island. *Wilson Bulletin* 82: 92–96.

Earl, J. P. 1950. Production of mallards on irrigated land in the Sacramento Valley, California. *Journal of Wildlife Management* 14: 332–342.

Figley, W. K., and L. W. VanDruff. 1982. *The Ecology of Urban Mallards.* Wildlife Society Monographs 81. The Wildlife Society, Washington, DC.

Fleskes, J. P., J. L. Yee, G. S. Yarris, M. R. Miller, and M. L. Casazza. 2007. Pintail and Mallard survival in California relative to habitat, abundance, and hunting. *Journal of Wildlife Management* 71: 2238–2248.

Gilmer, D. S., I. J. Ball, L. M. Cowardin, J. H. Riechmann, and J. R. Tester. 1975. Habitat use and home range of mallards breeding in Minnesota. *Journal of Wildlife Management* 39: 781–789.

Gilmer, D. S., R. E. Kirby, I. J. Ball, and J. H. Reichmann. 1977. Post-breeding activities of mallards and wood ducks in north-central Minnesota. *Journal of Wildlife Management* 41: 345–359.

Girard, G. L. 1941. The mallard, its management in western Montana. *Journal of Wildlife Management* 5: 233–259.

Heitmeyer, M. E. 1987. The prebasic moult and basic plumage of female mallards (*Anas platyrhynchos*). *Canadian Journal of Zoology* 65: 2248–2261.

Heitmeyer, M. E. 1988a. Body composition of female mallards in winter in relation to annual cycle events. *Condor* 90: 669–680.

Heitmeyer, M. E. 1988b. Protein costs of the prebasic molt of female mallards. *Condor* 90: 263–266.

Heitmeyer, M. E. 1988c. Changes in the visceral morphology of wintering female mallards (*Anas platyrhynchos*). *Canadian Journal of Zoology* 66: 2015–2018.

Heitmeyer, M. E. 2006. The importance of winter floods to mallards in the Mississippi Alluvial Valley. *Journal of Wildlife Management* 70: 101–110.

Heitmeyer, M. E., and L. H. Fredrickson. 1981. Do wetland conditions in the Mississippi Delta hardwoods influence mallard recruitment? *Transactions of the North American Wildlife and Natural Resources Conference* 46: 44–57.

Hestbeck, J. B. 1990. North-south gradient in survival rates in midcontinental populations of mallards. *Journal of Wildlife Management* 54: 206–210.

Heusmann, H. W. 1988. The role of parks in the range expansion of the mallard in the Northeast. Pp. 405–412 in M. W. Weller, ed. *Waterfowl in Winter.* University of Minnesota Press, Minneapolis.

Heusmann, H. W. 1991. The history and status of the mallard in the Atlantic Flyway. *Wildlife Society Bulletin* 19: 14–22.

Hoekman, S. T., T. S. Gabor, M. J. Petrie, R. Maher, H. R. Murkin, and M. S. Lindberg. 2006. Population dynamics of mallards breeding in agricultural environments in eastern Canada. *Journal of Wildlife Management* 70: 121–128.

Hoekman, S. T., T. S. Gabor, R. Maher, H. R. Murkin, and L. M. Armstrong. 2004. Factors affecting survival of mallard ducklings in southern Ontario. *Condor* 106: 485–495.

Hoekman, S. T., T. S. Gabor, R. Maher, H. R. Murkin, and M. S. Lindberg. 2006. Demographics of breeding female mallards in southern Ontario, Canada. *Journal of Wildlife Management* 70: 111–120.

Hori, J. 1963. Three-bird flights in the mallard. *Wildfowl Trust Annual Report* 14: 124–132.

Humburg, D. D., H. H. Prince, and R. A. Bishop. 1978. The social organization of a mallard population in northern Iowa. *Journal of Wildlife Management* 42: 72–80.

Johnson, O. W. 1961. Reproductive cycle of the mallard duck. *Condor* 63: 351–364.

Johnson, W. P., and F. C. Rohwer. 1998. Pairing chronology and agonistic behaviors of wintering green-winged teal and mallards. *Wilson Bulletin* 110: 311–315.

Johnson, W. P., and F. C. Rohwer. 2000. Foraging behavior of green-winged teal and mallards on tidal mudflats in Louisiana. *Wetlands* 20: 184–188.

Jorde, D. G., G. L. Krapu, R. D. Crawford. 1983. Feeding ecology of mallards wintering in Nebraska. *Journal of Wildlife Management* 47: 1044–1053.

Jorde, D. G., G. L. Krapu, R. D. Crawford, and M. A. Hay. 1984. Effects of weather on habitat selection and behavior of mallards wintering in Nebraska. *Condor* 86: 258–265.

Kaminski, R. M., and E. A. Gluessing. 1987. Density- and habitat-related recruitment in mallards. *Journal of Wildlife Management* 51: 141–148.

Kaminski, R. M., R. W. Alexander, and B. D. Leopold. 1993. Wood duck and mallard microhabitats in Mississippi hardwood bottomlands. *Journal of Wildlife Management* 57: 562–570.

Kirby, R. E., J. H. Reichmann, and L. M. Cowardin. 1985. Home range and habitat use of forest-dwelling mallards in Minnesota. *Wilson Bulletin* 97: 215–219.

Klint, T. 1980. Influence of male nuptial plumage on mate selection in the mallard (*Anas platyrhynchos*). *Animal Behaviour* 28: 1230–1238.

Krapu, G. L. 1981. The role of nutrient reserves in mallard reproduction. *Auk* 98: 29–38.

Krapu, G. L., L. G. Talent, and T. J. Dwyer. 1979. Marsh nesting by mallards. *Wildlife Society Bulletin* 7: 104–110.

Krapu, G. L., P. J. Pietz, D. A. Brandt, and R. R. Cox, Jr. 2000. Factors limiting mallard brood survival in prairie pothole landscapes. *Journal of Wildlife Management* 64: 553–561.

Krapu, G. L., P. J. Pietz, D. A. Brandt, and R. R. Cox, Jr. 2006. Mallard brood movements, wetland use, and duckling survival during and following a prairie drought. *Journal of Wildlife Management* 70: 1436–1444.

LaGrange, T. G., and J. J. Dinsmore. 1989. Habitat use by mallards during spring migration through central Iowa. *Journal of Wildlife Management* 53: 1076–1081.

Lebret, T. 1961. The pair formation in the annual cycle of the mallard, *Anas platyrhynchos* L. *Ardea* 49: 97–158.

Link, P. T., A. D. Afton, R. R. Cox, Jr., and B. E. Davis. 2011a. Daily movements of female mallards wintering in southwestern Louisiana. *Waterbirds* 34: 422–428.

Link, P. T., A. D. Afton, R. R. Cox, Jr., and B. E. Davis. 2011b. Use of habitats by female mallards wintering in southwestern Louisiana. *Waterbirds* 34: 429–438.

Losito, M. P., and G. A. Baldassarre. 1995. Wetland use by breeding and postbreeding mallards in the St. Lawrence River Valley. *Wilson Bulletin* 107: 55–63.

Losito, M. P., and G. A. Baldassarre. 1996. Pair-bond dissolution in mallards. *Auk* 117: 692–695.

Majewski, P., and P. Beszterda. 1990. Influence of nesting success on female homing in the mallard. *Journal of Wildlife Management* 54: 459–462.

Mauser, D. M., R. L. Jarvis, and D. S. Gilmer. 1994a. Movements and habitat use of mallard broods in northeastern California. *Journal of Wildlife Management* 58: 88–94.

Mauser, D. M., R. L. Jarvis, and D. S. Gilmer. 1994b. Survival of radio-marked mallard ducklings in northeastern California. *Journal of Wildlife Management* 58: 82–87.

McKinney, F. 1953. Incubation and hatching behaviour in the mallard. *Wildfowl Trust Annual Report* 5: 68–70.

Mjelstad, H., and M. Sætersdal. 1990. Reforming of resident mallard pairs *Anas platyrhynchos*, rule rather than exception? *Wildfowl* 41: 150–151.

Mulhern, J. H., T. D. Nudds, and B. R. Neal. 1985. Wetland selection by mallards and blue-winged teal. *Wilson Bulletin* 97: 473–485.

Munro, J. A. 1943. Studies of waterfowl in British Columbia: Mallard. *Canadian Journal of Research* 21: 223–260.

Nichols, J. D., and J. E. Hines. 1983. The relationship between harvest and survival rates of mallards: A straightforward approach with portioned data sets. *Journal of Wildlife Management* 47: 334–348.

Nichols, J. D., K. J. Reinecke, and J. E. Hines. 1983. Factors affecting the distribution of mallards wintering in the Mississippi alluvial valley. *Auk* 100: 932–946.

Nichols, J. D., R. S. Pospahala, and J. E. Hines. 1982. Breeding-ground habitat conditions and the survival of mallards. *Journal of Wildlife Management* 46: 80–87.

Nudds, T. D., M. W. Miller, and C. D. Ankney. 1996. Black ducks: Harvest, mallards, or habitat? *International Waterfowl Symposium* 7: 50–60.

Ogilvie, M. A. 1964. A nesting study of mallard in Berkeley New Decoy, Slimbridge. *Wildfowl Trust Annual Report* 15: 84–88.

Ohde, B. R., R. A. Bishop, and J. J. Dinsmore. 1983. Mallard reproduction in relation to sex ratios. *Journal of Wildlife Management* 47: 118–126.

Olsen, R. E., R. R. Cox, Jr., A. D. Afton, and C. D. Ankney. 2011. Diet and gut morphology of male mallards during winter in North Dakota. *Waterbirds* 34: 59–69.

Omland, K. E. 1996a. Female mallard mating preferences for multiple male ornaments, 1: Natural variation. *Behavioral Ecology and Sociobiology* 39: 353–360.

Omland, K. E. 1996b. Female mallard mating preferences for multiple male ornaments, 2: Experimental variation. *Behavioral Ecology and Sociobiology* 39: 361–366.

Pehrsson, O. 1979. Feeding behaviour, feeding habitat utilization, and feeding efficiency of mallard ducklings (*Anas platyrhynchos* L.) as guided by a domestic duck. *Viltrevy* 10: 193–218.

Pehrsson, O. 1984. Relationships of food to spatial and temporal breeding strategies of mallards in Sweden. *Journal of Wildlife Management* 48: 322–339.

Pietz, P. J., and D. A. Buhl. 1999. Behaviour patterns of mallard *Anas platyrhynchos* pairs and broods in Minnesota and North Dakota. *Wildfowl* 50: 101–122.

Pietz, P. J., G. L. Krapu, D. A. Buhl, and D. A. Brandt. 2000. Effects of water conditions on clutch size, egg volume, and hatching mass of mallards and gadwalls in the Prairie Pothole region. *Condor* 102: 936–940.

Pospahala, R. S., D. R. Anderson, and C. J. Henny. 1974. *Population Ecology of the Mallard, 2: Breeding Habitat Conditions, Size of the Breeding Populations, and Production Indices.* Resource Publication 115. U.S. Bureau of Sport Fisheries and Wildlife, Washington, DC.

Raitasuo, K. 1964. Social behaviour of the mallard in the course of the annual cycle. *Papers on Game Research* (*Riistatieteellisiä Julkaisuja*) 24: 1–62.

Raven, G. H., L. M. Armstrong, D. W. Howerter, and T. W. Arnold. 2007. Wetland selection by mallard broods in Canada's prairie parklands. *Journal of Wildlife Management* 71: 2527–2531.

Reinecke, K. J., C. W. Shaiffer, and D. Delnicki. 1987. Winter survival of female mallards in the lower Mississippi Valley. *Transactions of the North American Wildlife and Natural Resources Conference* 52: 258–263.

Reynolds, R. E., R. J. Blohm, J. D. Nichols, and J. E. Hines. 1995. Spring-summer survival rates of yearling versus adult female mallards. *Journal of Wildlife Management* 59: 691–696.

Reynolds, R. E., and J. R. Sauer. 1991. Changes in mallard breeding populations in relation to production and harvest rates. *Journal of Wildlife Management* 55: 483–487.

Rhymer, J. M. 1988. The effect of egg size variability on thermoregulation of mallard (*Anas platyrhynchos*) offspring and its implications for survival. *Oecologia* 75: 20–24.

Richardson, D. M., and R. M. Kaminski. 1992. Diet restriction, diet quality, and prebasic molt in female mallards. *Journal of Wildlife Management* 56: 531–539.

Ringelman, J. K., and L. D. Flake. 1980. Diurnal visibility and activity of wigeon-winged teal and mallard broods. *Journal of Wildlife Management* 44: 822–829.

Rotella, J. J., D. W. Howerter, T. P. Sankowski, and J. H. Devries. 1993. Nesting effort by wild mallards with 3 types of radio transmitters. *Journal of Wildlife Management* 57: 690–695.

Sayler, R. D., and M. A. Williams. 1997. Brood ecology of mallards and gadwalls nesting on islands in large reservoirs. *Journal of Wildlife Management* 61: 808–815.

Smith, G. W., and R. E. Reynolds. 1992. Hunting and mallard survival, 1979–88. *Journal of Wildlife Management* 56: 306–316.

Sugden, L. G., and E. A. Driver. 1980. Natural foods of mallards in Saskatchewan parklands during late summer and fall. *Journal of Wildlife Management* 44: 705–709.

Sugden, L. G., E. A. Driver, and M. C. S. Kingsley. 1981. Growth and energy consumption by captive mallards. *Canadian Journal of Zoology* 59: 1567–1570.

Swanson, G. A., M. I. Meyer, and V. A. Adomaitis. 1985. Foods consumed by breeding mallards on wetlands of south-central North Dakota. *Journal of Wildlife Management* 49: 197–203.

Swanson, G. A., T. L. Shaffer, J. F. Wolf, and F. B. Lee. 1986. Renesting characteristics of captive mallards in experimental ponds. *Journal of Wildlife Management* 50: 32–38.

Talent, L. G., G. L. Krapu, and R. L. Jarvis. 1982. Habitat use by mallard broods in south-central North Dakota. *Journal of Wildlife Management* 46: 629–635.

Talent, L. G., R. L. Jarvis, and G. L. Krapu. 1983. Survival of mallard broods in south-central North Dakota. *Condor* 85: 74–78.

Titman, R. D. 1983. Spacing and three-bird flights of mallards breeding in pothole habitat. *Canadian Journal of Zoology* 61: 839–847.

Titman, R. D., and J. K. Lowther. 1975. The breeding behavior of a crowded population of mallards. *Canadian Journal of Zoology* 53: 1270–1283.

Trost, R. E. 1987. Mallard survival and harvest rates: A reexamination of relationships. *Transactions of the North American Wildlife and Natural Resources Conference* 52: 264–284.

Turnbull, R. E., and G. A. Baldassarre. 1987. Activity budgets of mallards and American wigeon wintering in east-central Alabama. *Wilson Bulletin* 99: 457–464.

Weidmann, U., and J. Darley. 1971. The role of the female in the social display of mallards. *Animal Behaviour* 19: 287–298.

Whyte, R. J., G. A. Baldassarre, and E. G. Bolen. 1986. Winter condition of mallards on the southern High Plains of Texas. *Journal of Wildlife Management* 50: 52–57.

Wright, T. W. 1961. Winter foods of mallards in Arkansas. *Proceedings of the Annual Conference of the Southeastern Association of Game and Fish Commissioners* 13: 291–296.

Yarris, G. S., M. R. McLandress, and A. E. H. Perkins. 1994. Molt migration of postbreeding female mallards from Suisun Marsh, California. *Condor* 96: 36–45.

Yetter, A. P., J. D. Stafford, C. S. Hine, M. W. Bowyer, S. P. Havera, and M. M. Horath. 2009. Nesting biology of mallards in west-central Illinois. *Illinois Natural History Survey Bulletin* 39: 1–38.

Young, D. A., and D. A. Boag. 1981. A description of moult in male mallards. *Canadian Journal of Zoology* 59: 252–259.

Mexican, Florida, and Mottled Ducks

Aldrich, J. W., and K. P. Baer. 1970. Status and speciation in the Mexican duck (*Anas diazi*). *Wilson Bulletin* 82: 63–73.

Baker, O. E., III. 1983. Nesting and brood rearing habits of the mottled duck in the coastal marsh of Cameron Parish, Louisiana. MS thesis, Louisiana State University, Baton Rouge.

Ballard, B. M., M. T. Merendino, R. H. Terry, and T. C. Tacha. 2001. Estimating abundance of breeding mottled ducks in Texas. *Wildlife Society Bulletin* 29: 1186–1192.

Beckwith, S. L., and H. J. Hosford. 1955. The Florida duck in the vicinity of Lake Okeechobee, Glades County, Florida. *Proceedings of the Annual Conference of the Southeastern Association of Fish and Wildlife Agencies* 9: 188–201.

Beckwith, S. L., and H. J. Hosford. 1957. A report on seasonal food habits and life history notes of the Florida duck in the vicinity of Lake Okeechobee, Glades County, Florida. *American Midland Naturalist* 57: 461–473.

Bevill, W. V., Jr. 1970. Effects of supplemental stocking and habitat development on abundance of the Mexican duck. MS thesis, New Mexico State University, Las Cruces.

Bielefeld, R. R. 2002. *Habitat Use, Survival, and Movements of Florida Mottled Duck* (Anas fulvigula fulvigula) *Females Using the Upper St. Johns River Basin in Florida.* Final Report. Florida Fish and Wildlife Research Institute, Fish and Wildlife Conservation Commission, St. Petersburg, FL.

Bielefeld, R. R. 2008. *Mottled Duck Survey Report.* Florida Fish and Wildlife Research Institute, Fish and Wildlife Conservation Commission, St. Petersburg, FL.

Bielefeld, R. R. 2011. *Habitat Use, Movements, and Survival of Florida Mottled Duck* (Anas fulvigula fulvigula) *Females and Characterization of Wetlands Used by Dabbling Ducks in South Florida.* Florida Fish and Wildlife Research Institute, Fish and Wildlife Conservation Commission, St. Petersburg, FL.

Bielefeld, R. R., and R. R. Cox, Jr. 2006. Survival and cause-specific mortality of adult female mottled ducks in east-central Florida. *Wildlife Society Bulletin* 34: 388–394.

Callaghan, D. 2005. Mottled duck. Pp. 517–520 in J. Kear, ed. *Ducks, Geese, and Swans.* Vol. 2. Oxford University Press, Oxford, UK. 898 pp.

Chamberlain, E. B. 1960. *Florida Waterfowl Populations, Habitats, and Managements.* Florida Game and Fresh Water Fish Commission, Technical Bulletin No. 7. 62 pp.

Dugger, B. D., R. Finger, and S. L. Melvin. 2010. Nesting ecology of mottled ducks *Anas fulvigula* in interior Florida. *Wildfowl* 60: 95–105.

Durham, R. S., and A. D. Afton. 2003. Nest-site selection and success of mottled ducks on agricultural lands in southwest Louisiana. *Wildlife Society Bulletin* 31: 433–442.

Durham, R. S., and A. D. Afton. 2006. Breeding biology of mottled ducks on agricultural lands in southwestern Louisiana. *Southeastern Naturalist* 5: 311–316.

Engeling, G. A. 1949. The mottled duck, a determined nester. *Texas Game and Fish* 7(8): 6–7, 19.

Engeling, G. A. 1950. The nesting habits of the mottled duck in Wharton, Fort Bend, and Brazoria Counties, Texas, with notes on molting and movements. MS thesis, Texas A&M University, College Station.

Engeling, G. A. 1951. Mottled duck movements in Texas. *Texas Game and Fish* 9(5): 2–5.

Finger, R. S., B. M. Ballard, M. T. Merendino, J. P. Hurst, D. S. Lobpries, and A. M. Fedynich. 2003. *Habitat Use, Movements, and Survival of Female Mottled Ducks and Ducklings during Brood Rearing: Final Report.* Texas Parks and Wildlife Department, Austin.

Fogarty, M. J., and D. E. LaHart. 1971. Florida duck movements. *Proceedings of the Southeastern Association of Fish and Wildlife Agencies* 25: 191–202.

Grand, J. B. 1988. Habitat selection and social structure of mottled ducks in a Texas coastal marsh. PhD dissertation, Texas A&M University, College Station.

Grand, J. B. 1992. Breeding chronology of mottled ducks in a Texas coastal marsh. *Journal of Field Ornithology* 63: 195–202.

Gray, P. N. 1993. The biology of a southern mallard, Florida's mottled duck. PhD dissertation, University of Florida, Gainesville.

Haukos, D. A. 2009. *Status of the Western Gulf Coast Population of Mottled Ducks.* US Fish and Wildlife Service, Albuquerque.

Haukos, D. A., S. Martinez, and J. Hetzel. 2010. Characteristics of ponds used by breeding mottled ducks on the Chenier Plain of the Texas Gulf Coast. *Journal of Fish and Wildlife Management* 1: 93–101.

Holbrook, R. S. 1997. Ecology of nesting mottled ducks at the Atchafalaya River Delta, Louisiana. MS thesis, Louisiana State University, Baton Rouge.

Hubbard, J. P. 1977. *The Biological and Taxonomic Status of the Mexican Duck,* New Mexico Department of Game and Fish Bulletin 16, pp. 1–56.

Huey, W. S. 1961. Comparison of female mallard with female New Mexican duck. *Auk* 78: 428–431.

Hyde, R. A. 1958. *Florida Waterfowl Band Recoveries, 1920–1957*. Florida Game and Fresh Water Fish Commission, Report on F. A. Project W-19-R, pp. 1–57.

Johnson, F. A. 2009. *Variation in Population Growth Rates of Mottled Ducks in Texas and Louisiana*. US Geological Survey Administrative Report. US Fish and Wildlife Service, Reston, VA.

Johnson, F. A., and F. Montalbano, III. 1984. Selection of plant communities by wintering waterfowl on Lake Okeechobee, Florida. *Journal of Wildlife Management* 48: 174–178.

Johnson, F. A., D. H. Brakhage, R. E. Turnbull, and F. Montalbano, III. 1995. Variation in band-recovery and survival rates of mottled ducks in Florida. *Proceedings of the Southeastern Association of Fish and Wildlife Agencies* 49: 594–606.

Johnson, F. A., F. Montalbano, III, and T. C. Hines. 1984. Population dynamics and status of the mottled duck in Florida. *Journal of Wildlife Management* 48: 1137–1143.

Johnson, F. A., F. Montalbano, III, J. D. Truitt, and D. R. Eggeman. 1991. Distribution, abundance, and habitat use by mottled ducks in Florida. *Journal of Wildlife Management* 55: 476–482.

Johnson, W. P., F. C. Rohwer, and M. Carloss. 1996. Evidence of nest parasitism in mottled ducks. *Wilson Bulletin* 108: 187–189.

Johnson, W. P., R. S. Holbrook, and F. C. Rohwer. 2002. Nesting chronology, clutch size, and egg size in the mottled duck. *Wildfowl* 53: 155–166.

Lindsey, A. A. 1946. The nesting of the New Mexican duck. *Auk* 63: 483–492.

McHenry, M. G. 1968. Mottled ducks in Kansas. *Wilson Bulletin* 80: 229–230.

McKenzie, P. M., P. J. Zwank, and E. B. Moser. 1988. Mottled duck population trends based on analyses of Christmas Bird Count data. *American Birds* 42: 512–516.

Merendino, M. T., D. S. Lobpries, J. E. Neaville, J. D. Ortega, and W. P. Johnson. 2005. Regional differences and long-term trends in lead exposure in mottled ducks. *Wildlife Society Bulletin* 33: 1002–1008.

Montalbano, F., III. 1980. Summer use of two central Florida phosphate settling ponds by Florida ducks. *Proceedings of the Southeastern Association of Fish and Wildlife Agencies* 34: 584–590.

Neaville, J. 1996. *Mottled Duck Breeding Pair Survey Status Report*. Administrative Report, US Fish and Wildlife Service, Washington, DC.

Nymeyer, L. A. 1975. *The Mexican Duck in Southcentral New Mexico: Distribution, Abundance, Habitat*. Unpublished report, New Mexico Game and Fish Department, Albuquerque.

O'Brien, G. P. 1975. *A Study of the Mexican Duck* (Anas diazi) *in Southeastern Arizona*. Special Report 5. Arizona Game and Fish Department, Phoenix.

Ohlendorf, H. M., and R. F. Patton. 1971. Nesting record of the Mexican duck (*Anas diazi*) in Texas. *Wilson Bulletin* 83: 97.

Paulus, S. L. 1984. Behavioral ecology of mottled ducks in Louisiana. PhD dissertation, Auburn University, Auburn, AL.

Paulus, S. L. 1988. Social behavior and pairing chronology of mottled ducks during autumn and winter in Louisiana. Pp. 59–70 in M. W. Weller, ed. *Waterfowl in Winter*. University of Minnesota Press, Minneapolis.

Pérez-Arteaga, A., K. J. Gaston, and M. Kershaw. 2002. Population trends and priority conservation sites for Mexican duck *Anas diazi*. *Bird Conservation International* 12: 35–52.

Rigby, E. A. 2008. Recruitment of mottled ducks (*Anas fulvigula*) on the upper Texas Gulf Coast. MS thesis, Texas Tech University, Lubbock.

Scott, N. J., Jr., and R. P. Reynolds. 1984. Phenotypic variation of the Mexican duck (*Anas platyrhynchos diazi*) in Mexico. *Condor* 86: 266–274.

Sincock, J. L. 1957. *Florida Waterfowl Investigations*. Federal Aid in Wildlife Restoration Project W-19-R. Florida Game and Fresh Water Fish Commission, St. Petersburg.

Singleton, J. R. 1953. *Texas Coastal Waterfowl Survey*. Austin, TX: Texas Game and Fish Commission, F.A. Report, Series 11. 128 pp.

Singleton, J. R. 1968. Texas' mistaken mallards. *Texas Parks and Wildlife* 26: 8–11.

Stevenson, H. M., and B. H. Anderson. 1994. *The Birdlife of Florida*. University Press of Florida, Gainesville.

Stieglitz, W. O. 1972. Food habits of the Florida duck. *Journal of Wildlife Management* 36: 422–428.

Stieglitz, W. O., and C. T. Wilson. 1968. Breeding biology of the Florida duck. *Journal of Wildlife Management* 32: 921–934.

Stutzenbaker, C. D. 1988. *The Mottled Duck: Its Life History, Ecology, and Management.* Texas Parks amd Wildlife Department, Austin.

Swarbrick, B. M. 1975. Ecology of the Mexican duck in the Sulphur Springs Valley of Arizona. MS thesis, University of Arizona, Tucson.

Walters, N. F. 2000. Nesting activities of mottled ducks in the Mississippi River Delta. MS thesis, Louisiana State University, Baton Rouge.

Walters, N. F., F. C. Rohwer, and J. O. Harris. 2001. Nest success and nesting habitats of mottled ducks on the Mississippi River Delta in Louisiana. *Proceedings of the Southeastern Association of Fish and Wildlife Agencies* 55: 498–505.

Webster, R. E. 2006. The status of mottled duck (*Anas fulvigula*) in Arizona. *Arizona Birds Online* 2: 6–9.

Weeks, J. L. 1969. Breeding behavior of mottled ducks in Louisiana. MS thesis, Louisiana State University, Baton Rouge.

Weng, G.-J. 2006. Ecology and population genetics of mottled ducks within the South Atlantic Coastal Zone. PhD dissertation, University of Georgia, Athens.

Williams, S. O., III. 1980. The Mexican duck in Mexico: Natural history, distribution, and population status. PhD dissertation, Colorado State University, Fort Collins.

Wilson, B. C. 2007. *Gulf Coast Joint Venture: Mottled Duck Conservation Plan.* Mottled Duck Working Group, Gulf Coast Joint Venture, North American Waterfowl Management Plan, Albuquerque, NM.

Young, G. 2005. Mexican duck. Pp. 521–523 in J. Kear, ed. *Ducks, Geese, and Swans.* Vol. 2. Oxford University Press, Oxford, UK.

Zwank, P. J., P. M. McKenzie, and E. B. Moser. 1989. Mottled duck habitat use and density indices in agricultural lands. *Journal of Wildlife Management* 53: 110–114.

American Black Duck

Barske, P., ed. 1968. *The Black Duck: Evaluation, Management, and Research: A Symposium.* Atlantic Waterfowl Council and Wildlife Management Institute, Boston, MA.

Bordage, D., C. Lepage, and S. Orichefsky. 2003. *2003 Black Duck Joint Venture Helicopter Survey—Québec.* Environment Canada and Canadian Wildlife Service, Québec Region, Sainte-Foy.

Brodsky, L. M., and P. J. Weatherhead. 1985a. Time and energy constraints on courtship in wintering American black ducks. *Condor* 87: 33–36.

Brodsky, L. M., and P. J. Weatherhead. 1985b. Diving by wintering black ducks: an assessment of atypical foraging. *Wildfowl* 36: 72–76.

Brook, R. W., R. R. Kenyon, K. F. Abraham, D. L. Fronczak, and C. J. Davies. 2009. Evidence for black duck winter distribution change. *Journal of Wildlife Management* 73: 98–103.

Costanzo, G. R. 1988. Wintering ecology of black ducks along coastal New Jersey. PhD dissertation, Cornell University, Ithaca, NY.

Coulter, M. W., and H. E. Mendall. 1968. Habitat and breeding ecology: Northeastern states. Pp. 90–101 in P. Barske, ed. *The Black Duck: Evaluation, Management, and Research: A Symposium.* Atlantic Waterfowl Council and Wildlife Management Institute, Boston, MA.

Diefenbach, D. R., and R. B. Owen, Jr. 1989. A model of habitat use by breeding American black ducks. *Journal of Wildlife Management* 53: 383–389.

Diefenbach, D. R., J. D. Nichols, and J. E. Hines. 1988. Distribution patterns during winter and fidelity to wintering areas of American black ducks. *Canadian Journal of Zoology* 66: 1506–1513.

Francis, C. M., J. R. Sauer, and J. R. Serie. 1998. Effect of restrictive harvest regulations on survival and recovery rates of American black ducks. *Journal of Wildlife Management* 62: 1544–1557.

Frazer, C. 1988. The ecology of post-fledging American black ducks in eastern Maine. MS thesis, University of Maine, Orono.

Frazer, C., J. R. Longcore, and D. G. McAuley. 1990a. Habitat use by postfledging American black ducks in Maine and New Brunswick. *Journal of Wildlife Management* 54: 541–549.

Frazer, C., J. R. Longcore, and D. G. McAuley. 1990b. Home range and movements of postfledging American black ducks in eastern Maine. *Canadian Journal of Zoology* 68: 1288–1291.

Haramis, G. M., D. S. Chu, A. W. Diamond, and F. L. Filion. 1987. Acid rain effects on waterfowl: Use of black duck broods to assess food resources of experimentally acidified wetlands. Pp. 173–181 in A. W. Diamond and F. L. Filion, eds. *The Value of Birds*. Technical Publication 6. International Council for Bird Preservation, Cambridge, UK.

Hartman, F. E. 1963. Estuarine wintering habitat for black ducks. *Journal of Wildlife Management* 27: 339–347.

Heusmann, H. W., W. W. Blandin, and P. R. Pekkala. 1979. Hand-reared black ducks and elevated nesting cylinders. *Transactions of the Northeast Section of The Wildlife Society* 36: 138–144.

Hickey, T. E., and R. D. Titman. 1983. Diurnal activity budgets of black ducks during their annual cycle in Prince Edward Island. *Canadian Journal of Zoology* 61: 743–749.

Jorde, D. J., and R. B. Owen. 1990. Foods of black ducks, *Anas rubripes*, wintering in marine habitat of Maine. *Canadian Field-Naturalist* 104: 300–302.

Krementz, D. G. 1991. American black duck, *Anas rubripes*. Pp. 16.1–16.7 in S. L. Funderburk, J. A. Mihursky, S. J. Jordan, and D. Riley, eds. *Habitat Requirements for Chesapeake Bay Living Resources*. 2nd ed. Living Resources Subcommittee, Chesapeake Research Consortium, Solomons, MD.

Krementz, D. G., J. E. Hines, P. O. Corr, and R. B. Owen, Jr. 1989. The relationship between body mass and annual survival in American black ducks. *Ornis Scandinavica* 20: 81–85.

Krementz, D. G., M. J. Conroy, J. E. Hines, and H. F. Percival. 1988. The effects of hunting on survival rates of American black ducks. *Journal of Wildlife Management* 52: 214–226.

Longcore, J. R., and D. G. McAuley. 2004. Extraordinary size and survival of American black duck, *Anas rubripes*, broods. *Canadian Field-Naturalist* 118: 129–131.

Longcore, J. R., and J. P. Gibbs. 1988. Distribution and numbers of American black ducks along the Maine coast during the severe winter of 1980–1981. Pp. 377–389 in M. W. Weller, ed. *Waterfowl in Winter*. University of Minnesota Press, Minneapolis.

Longcore, J. R., D. G. McAuley, and C. Frazer. 1991. Survival of postfledging female American black ducks. *Journal of Wildlife Management* 55: 573–580.

Longcore, J. R., D. G. McAuley, D. A. Clugston, C. M. Bunck, J.-F. Giroux, C. Ouellet, G. R. Parker, P. Dupuis, D. B. Stotts, and J. R. Goldsberry. 2000b. Survival of American black ducks radiomarked in Québec, Nova Scotia, and Vermont. *Journal of Wildlife Management* 64: 238–252.

Martinson, R. K., A. D. Geis, and R. I. Smith. 1968. Black duck harvest and population dynamics in eastern Canada and the Atlantic Flyway. Pp. 21–52 in P. Barske, ed. *The Black Duck: Evaluation, Management, and Research: A Symposium*. Atlantic Waterfowl Council and Wildlife Management Institute, Boston, MA.

McGilvrey, F. 1971. Conditioning black ducks to nest in elevated cylinders. *Transactions of the Northeast Section of the Wildlife Society* 28: 213–220.

Mendall, H. L. 1949. Food habits in relation to black duck management in Maine. *Journal of Wildlife Management* 13: 64–101.

Merendino, M. T., and C. D. Ankeny. 1994. Habitat use by mallards and American black ducks breeding in central Ontario. *Condor* 96: 411–421.

Morton, J. M., A. C. Fowler, and R. L. Kirkpatrick. 1989a. Time and energy budgets of American black ducks wintering at Chincoteague, Virginia. *Journal of Wildlife Management* 59: 401–410.

Morton, J. M., R. L. Kirkpatrick, and M. R. Vaughn. 1990. Changes in body composition of American black ducks wintering at Chincoteague, Virginia. *Condor* 92: 598–605.

Morton, J. M., R. L. Kirkpatrick, M. R. Vaughn, and D. F. Stauffer. 1989b. Habitat use and movements of American black ducks in winter. *Journal of Wildlife Management* 53: 390–400.

Murray, L. H. 1958. The black duck in Saskatchewan. *Blue Jay* 16: 109–111.

Nichols, J. D. 1991. Science, population ecology, and the management of the American black duck. *Journal of Wildlife Management* 55: 790–799.

Parker, G. R. 1991. Survival of juvenile American black ducks on a managed wetland in New Brunswick. *Journal of Wildlife Management* 55: 466–470.

Plattner, D. M., M. W. Eicholz, and T. Yerkes. 2010. Food resources for wintering and spring staging black ducks. *Journal of Wildlife Management* 74: 1554–1558.

Rattner, B. A., G. M. Haramis, D. S. Chu, and C. M. Bunck. 1987. Growth and physiological condition of black ducks reared on acidified wetlands. *Canadian Journal of Zoology* 65: 2953–2958.

Reed, A. 1968. Habitat and breeding ecology: Eastern Canada. Pp. 57–89 in P. Barske, ed. *The Black Duck: Evaluation, Management, and Research: A Symposium.* Atlantic Waterfowl Council and Wildlife Management Institute, Boston, MA.

Reed, A. 1970. The breeding ecology of the black duck in the St. Lawrence Estuary. PhD dissertation, Laval University, Québec City, QC.

Reed, A. 1975. Reproductive output of black ducks in the St. Lawrence Estuary. *Journal of Wildlife Management* 39: 243–255.

Reinecke, K. J. 1979. Feeding ecology and development of juvenile black ducks in Maine. *Auk* 96: 737–745.

Reinecke, K. J., and R. B. Owen, Jr. 1980. Food use and nutrition of black ducks nesting in Maine. *Journal of Wildlife Management* 44:549–558.

Reinecke, K. J., T. L. Stone, and R. B. Own, Jr. 1982. Seasonal carcass composition and energy balance of female black ducks in Maine. *Condor* 84: 420–426.

Ringelman, J. K., and J. R. Longcore. 1982a. Movements and wetland selection by brood-rearing black ducks. *Journal of Wildlife Management* 46: 615–621.

Ringelman, J. K., and J. R. Longcore. 1982b. Survival of juvenile black ducks during brood rearing. *Journal of Wildlife Management* 46: 622–628.

Ringelman, J. K., and J. R. Longcore. 1983. Survival of female black duck, *Anas rubripes*, during the breeding season. *Canadian Field-Naturalist* 97: 62–65.

Ringelman, J. K., J. R. Longcore, and R. B. Owen, Jr. 1982a. Breeding habitat selection and home range of radio-marked black ducks (*Anas rubripes*) in Maine. *Canadian Journal of Zoology* 60: 241–248.

Ringelman, J. K., J. R. Longcore, and R. B. Owen, Jr. 1982b. Nest and brood attentiveness in female black ducks. *Condor* 84: 110–116.

Samuel, M. D., and E. F. Bowers. 2000. Lead exposure in American black ducks after implementation of non-toxic shot. *Journal of Wildlife Management* 64: 947–953.

Sanders, M. A., D. L. Combs, M. J. Conroy, and J. F. Hopper. 1995. Distribution patterns of American black ducks wintering in Tennessee. *Proceedings of the Southeastern Association of Fish and Wildlife Agencies* 49: 607–617.

Seymour, N. R. 1984. Activity of black ducks nesting along streams in northeastern Nova Scotia. *Wildfowl* 35: 143–150.

Seymour, N. R. 1991. Philopatry in male and female American black ducks. *Condor* 93: 189–191.

Seymour, N. R, and R. D. Titman. 1978. Changes in activity patterns, agonistic behavior, and territoriality of black ducks (*Anas rubripes*) during the breeding season in a Nova Scotia tidal marsh. *Canadian Journal of Zoology* 56: 1773–1785.

Seymour, N. R., and W. Jackson. 1996. Habitat-related variation in movements and fledging success of American black duck broods in northeastern Nova Scotia. *Canadian Journal of Zoology* 74: 1158–1164.

Stotts, V. 1935. Black duck breeding study ends in the Kent Island area. *Maryland Tidewater News* 12(4): 1, 4.

Stotts, V. 1957. The black duck (*Anas rubripes*) in the Upper Chesapeake Bay. *Proceedings 10th Annual Conference, Southeastern Association of Game and Fish Commissioners*, pp. 234–242.

Stotts, V. 1958. The time of formation of pairs in black ducks, *North American Wildlife Conference Transactions* 23: 192–197.

Stotts, V. 1968. Habitat and breeding ecology: East central United States. Pp. 102–112 in P. Barske, ed. *The Black Duck: Evaluation, Management, and Research: A Symposium.* Atlantic Waterfowl Council and Wildlife Management Institute, Boston, MA.

Stotts, V. D. 1987. *A Survey of Breeding American Black Ducks in the Eastern Bay Region of Maryland in 1986.* Report for contract 14-16-005-86-017. US Fish and Wildlife Service, Washington, DC.

Stotts, V. D., and D. E. Davis. 1960. The black duck in the Chesapeake Bay of Maryland: Breeding behavior and biology. *Chesapeake Science* 1: 127–154.

Wright, B. S. 1954. *High Tide and an East Wind: The Story of the Black Duck.* Management Institute, Washington, DC, and Stackpole, Harrisburg, PA.

Zimpfer, N. L., and M. J. Conroy. 2006. Modeling movement and fidelity of American black ducks. *Journal of Wildlife Management* 70: 1770–1777.

Mallard–Black Duck Interactions and Genetics

Ankney, C. D., D. G. Dennis, and R. C. Bailey. 1987. Increasing mallards, decreasing black ducks: Coincidence or cause and effect? *Journal of Wildlife Management* 51: 523–529.

Ankney, C. D., D. G. Dennis, L. N. Wishard, and J. E. Seeb. 1986. Low genetic variation between black ducks and mallards. *Auk* 103: 701–709.

Avise, J. C., D. C. Ankey, and W. S. Nelson. 1990. Mitochondrial gene trees and the evolutionary relationship of mallard and black ducks. *Evolution* 44: 1109–1119.

Barclay, J. S. 1970. Ecological aspects of defensive behavior in breeding mallards and black ducks. PhD dissertation, Ohio State University, Columbus.

Barnes, G. G., and T. D. Nudds. 1991. Salt tolerance in American black ducks, mallards, and their F-1 hybrids. *Auk* 108: 89–98.

Bélanger, L., and D. Lehoux. 1994. Use of a tidal saltmarsh and coastal impoundments by sympatric breeding and staging American black ducks, *Anas rubripes*, and mallards, *A. platyrhynchos*. *Canadian Field-Naturalist* 108: 311–317.

Bellrose, F. C., and R. D. Crompton. 1970. Migrational behavior of mallards and black ducks as determined from banding. *Illinois Natural History Survey Bulletin* 30: 167–234.

Brodsky, L. M., and P. J. Weatherhead. 1984. Behavioral and ecological factors contributing to American black duck–mallard hybridization. *Journal of Wildlife Management* 48: 846–852.

Brodsky, L. M., P. J. Weatherhead, and D. G. Dennis. 1988. The influence of male dominance on social interactions in black ducks and mallards. *Animal Behaviour* 36: 1371–1378.

Brook, R. W., R. R. Kenyon, K. F. Abraham, D. L. Fronczak, and C. J. Davies. 2009. Evidence for black duck winter distribution change. *Journal of Wildlife Management* 73: 98–103.

Conroy, M. J., G. G. Barnes, R. W. Bethke, and T. D. Nudds. 1989. Increasing mallards, decreasing American black ducks—No evidence for cause and effect: A comment. *Journal of Wildlife Management* 53: 1065–1071.

Coulter, M. W., and W. R. Miller. 1968. *Nesting Biology of Black Ducks and Mallards in Northern New England*. Bulletin 68-2. Vermont Fish and Game Department, Montpelier.

Cowardin, L. M., G. E. Cummings, and P. B. Reed, Jr. 1967. Stump and tree nesting by mallards and black ducks. *Journal of Wildlife Management* 31: 229–235.

Diefenbach, D. R., J. D. Nichols, and J. E. Hines. 1988b. Distribution patterns of American black duck and mallard winter band recoveries. *Journal of Wildlife Management* 52: 704–710.

Dwyer, C. P., and G. A. Baldassarre. 1993. Survival and nest success of sympatric female mallards, *Anas platyrhynchos*, and American black ducks, *A. rubripes*, breeding in a forested environment. *Canadian Field-Naturalist* 107: 213–216.

Dwyer, C. P., and G. A. Baldassarre. 1994. Habitat use by sympatric female mallards and American black ducks breeding in a forested environment. *Canadian Journal of Zoology* 72: 1538–1542.

Goodwin, C. E. 1956. Black duck and mallard populations in the Toronto area. *Ontario Field Biologist* 10: 7–18.

Grand, J. B. 1992. Breeding chronology of mottled ducks in a Texas coastal marsh. *J. Field Ornithology* 62: 195–202.

Heusmann, H. W. 1974. Mallard–black duck relationships in the Northeast. *Wildlife Society Bulletin* 2: 171–177.

Hepp, G. R., J. M. Novak, K. T. Scribner, and P. W. Stangel.1988. Genetic distance and hybridization of black ducks and mallards: A morph of a different color? *Auk* 105: 804–807.

Johnsgard, P. A. 1960. A quantitative study of sexual behavior of mallards and black ducks. *Wilson Bulletin* 72: 133–155.

Johnsgard, P. A. 1961. Wintering distribution changes in mallards and black ducks. *American Midland Naturalist* 66: 477–484. http://digitalcommons.unl.edu/biosciornithology/72/

Johnsgard, P. A. 1967. Sympatry changes and hybridization incidence in mallards and black ducks. *American Midland Naturalist* 77: 51–63.

Johnsgard, P. A., and R. DiSilvestro. 1976. Seventy-five years of changes in mallard–black duck ratios in eastern North America. *American Birds* 30: 904–908.

Kirby, R. E., A. Reed, P. Dupuis, H. H. Obrecht, III, and W. J. Quist. 2000. *Description and Identification of American Black Duck,*

Mallard, and Hybrid Wing Plumage. Biological Science Report USGS/BRD/BSR-2000-0002. Biological Resources Division, Northern Prairie Wildlife Research Center, US Geological Survey, Jamestown, ND.

Kirby, R. E., G. A. Sargeant, and D. Shutler. 2004. Haldane's rule and American black duck × mallard hybridization. *Canadian Journal of Zoology* 82: 1827–1831.

Krementz, D. G., and G. W. Pendleton. 1991. Movements and survival of American black duck and mallard broods on Chesapeake Bay. *Proceedings of the Southeastern Association of Fish and Wildlife Agencies* 45: 156–166.

Longcore, J. R., D. A. Clugston, and D. G. McAuley. 1998. Brood sizes of sympatric American black ducks and mallards in Maine. *Journal of Wildlife Management* 62: 142–151.

Longcore, J. R., P. O. Corr, and D. G. McAuley. 1987. Black duck–mallard interactions on breeding areas in Maine. *Transactions of the Northeast Section of the Wildlife Society* 44: 16–132.

Maisonneuve, C., A. Desrosiers, and R. McNicoll. 2000. Habitat use, movements, and survival of American black duck, *Anas rubripes*, and mallard, *A. platyrhynchos*, broods in agricultural landscapes of southern Québec. *Canadian Field-Naturalist* 114: 201–210.

Maisonneuve, C., L. Bélanger, D. Bordage, J. Benoit, G. Marcelle, J. Beaulieu, S. Gabor, and B. Filion. 2006. American black duck and mallard breeding distribution and habitat relationships along a forest-agriculture gradient in southern Québec. *Journal of Wildlife Management* 70: 450–459.

Maisonneuve, C., R. McNicoll, and A. Desrosiers. 2000. Comparative productivity of American black ducks and mallards nesting in agricultural landscapes of southern Québec. *Waterbirds* 23: 378–387.

Mank, J. E., J. E. Carlson, and M. C. Brittingham. 2004. A century of hybridization: Decreasing genetic distance between American black ducks and mallards. *Conservation Genetics* 5: 395–403.

McAuley, D. G., D. A. Clugston, and J. L. Longcore. 1998. Outcome of aggressive interactions between American black ducks and mallards during the breeding season. *Journal of Wildlife Management* 62: 134–141.

Merendino, M. T., and C. D. Ankeny. 1994. Habitat use by mallards and American black ducks breeding in central Ontario. *Condor* 96: 411–421.

Nummi., P. 1993. Food-niche relationships of sympatric mallards and green-winged teal. *Canadian Journal of Zoology* 71: 49–55.

Petrie, S. A., R. D. Drobney, and D. T. Sears. 2000. Mallard and black duck breeding parameters in New Brunswick: A test of the reproductive hypothesis. *Journal of Wildlife Management* 64: 832–838.

Phillips, J. C. 1912. A reconsideration of the American black duck with special reference to certain variations. *Auk* 29: 295–306.

Seymour, N. R. 1990. Forced copulation in sympatric American black ducks and mallards in Nova Scotia. *Canadian Journal of Zoology* 68: 1691–1696.

Other Mallard Hybrids and *Anas* Genetics

Braithwaite, L. W., and Miller, B. 1975. The mallard, *Anas platyrhynchos*, and mallard–black duck, *Anas superciliosa rogersi*, hybridization. *Australian Wildlife Research* 2: 47–61.

Browne, R. A., C. R. Griffin, P. R. Chang, M. Hubley, and A. E. Martin. 1993. Genetic divergence among populations of the Hawaiian duck, Laysan duck, and mallard. *Auk* 110: 49–56.

Gillespie, G. D. 1985. Hybridization, introgression, and morphometric differentiation between mallard (*Anas platyrhynchos*) and grey duck (*Anas superciliosa*) in Otago, New Zealand. *Auk* 102: 459–469.

Gonzalez, J., H. Düttmann, and M. Wink. 2009. Phylogenetic relationships based on two mitochondrial genes and hybridization patterns in Anatidae. *Journal of Zoology* 279: 310–318.

Haddon, M. 1984. A re-analysis of hybridization between mallard (*Anas platyrhynchos*) and grey ducks in New Zealand. *Auk* 101: 190–291.

Hubbard, J. P. 1977. *The Biological and Taxonomic Status of the Mexican Duck.* New Mexico Department of Game and Fish Bulletin 16, New Mexico Department of Game and Fish, Santa Fe.

Johnsgard, P. A 1959. Evolutionary relationships among the mallards. PhD dissertation, Cornell University, Ithaca, NY.

Johnsgard, P. A. 1961. Evolutionary relationships among the North American mallards. *Auk* 78: 1–43. http://digitalcommons. unl.edu/biosciornithology/62/

Johnson, K. P., and M. D. Sorenson.1999. Phylogeny and biogeography of dabbling ducks (genus *Anas*): a comparison of molecular and morphological evidence. *Auk* 116: 792–805.

Kulikova, I. V., Y. N. Zhuravlev, and K. G. McCracken. 2004. Asymmetric hybridization and sex-biased gene flow between eastern spot-billed ducks (*Anas zonorhyncha*) and mallards (*A. platyrhynchos*) in the Russian Far East. *Auk* 121: 930–949.

Lavretsky, P. 2008. Phylogenetics, population genetics, and evolution of the mallard complex. PhD dissertation, Wright State University, Dayton, OH.

Lavretsky, P., B. E. Hernández-Baños, and J. L. Peters 2014. Rapid radiation and hybridization contribute to weak differentiation and hinder phylogenetic inferences in the New World mallard complex (*Anas* spp.). *Auk* 131: 524–538.

Lockwood, M. W., and B. Freeman. 2004. *The TOS Handbook of Texas Birds*. Texas A&M University Press, College Station.

McCracken, K. G., W. P. Johnson, and F. H. Sheldon. 2001. Molecular population genetics, phylogeography, and conservation biology of the mottled duck (*Anas fulvigula*). *Conservation Genetics* 2: 87–102.

Nelson, D. 1980. A mallard × mottled duck hybrid. *Wilson Bulletin* 92: 527–529.

Phillips, J. C. 1915. Experimental studies of hybridization among ducks and pheasants. *Journal of Experimental Biology* 18: 69–144.

Phillips, J. C. 1921. A further report on species crosses in birds. *Genetics*. 6: 366–383.

Rhymer, J. M. 2006. Extinction by hybridization and introgression in anatine ducks. *Acta Zoologica Sinica* 52: 583–586.

Rhymer, J. M., and D. Simberloff. 1996. Extinction by hybridization and introgression. *Annual Review of Ecology and Systematics*. 27: 83–109. doi: 10.1146/annurev.ecolsys.27.1.83

Weeks, J. L. 1999. Breeding behavior of mottled ducks in Louisiana. MS thesis, Louisiana State University, Baton Rouge.

Williams, C. L., A. M. Fedynich, D. B. Pence, and O. E. Rhodes. 2005. Evaluation of allozyme and microsatellite variation in Texas and Florida mottled ducks. *Condor* 107: 155–161.

Williams, C. L., R. C. Brust, T. T. Fendley, G. R. Tiller, and O. E. Rhodes. 2005. A comparison of hybridization between mottled ducks (*Anas fulvigula*) and mallards (*A. platyrhynchos*) in Florida and South Carolina using microsatellite DNA analysis. *Conservation Genetics* 6: 445–453.

White-cheeked Pintail

Marchant, S.1960. The breeding of some S.W. Ecuadorean birds. *Ibis* 100: 349–382.

Sorenson, L. G. 1990. Breeding behavior and ecology of a sedentary tropical duck: The white-cheeked pintail. PhD dissertation, University of Minnesota, Minneapolis.

Sorenson, L. G. 1991. Mating systems of tropical and southern hemisphere dabbling ducks. *International Ornithological Congress* 20: 851–859.

Sorenson, L. G. 1992. Variable mating system of a sedentary tropical duck: The white-cheeked pintail (*Anas bahamensis bahamensis*). *Auk* 109: 277–292.

Sorenson, L. G. 1994. Forced extra-pair copulation in the white-cheeked pintail: Male tactics and female responses. *Condor* 96: 156–159.

Sorenson, L. G. 2005. White-cheeked pintail. Pp. 583–587 in J. Kear, ed. *Ducks, Geese, and Swans*. Vol. 2. Oxford University Press, Oxford, UK.

Sorenson, L. G., B. L. Woodworth, L. M. Ruttan, and F. McKinney 1992. Serial monogamy and double-brooding in the white-cheeked (Bahama) pintail, *Anas behamensis*. *Wildfowl* 42: 156–159.

Weller, M. 1968. Notes on some Argentine anatids. *Wilson Bulletin* 80: 189–212.

Northern Pintail

Baldassarre, G. A., and E. G. Bolen. 1986. Body weight and aspects of pairing chronology of green-winged teal and northern pintails wintering on the southern High Plains of Texas. *Southwestern Naturalist* 31: 361–366.

Ballard, B. M., J. E. Thompson, and M. J. Petrie. 2006. Carcass composition and digestive-tract dynamics of northern pintails wintering along the lower Texas Coast. *Journal of Wildlife Management* 70: 1316–1324.

Ballard, B. M., J. E. Thompson, M. J. Petrie, M. Checkett, and D. G. Hewitt. 2004. Diet and nutrition of northern pintails wintering along the southern coast of Texas. *Journal of Wildlife Management* 68: 371–382.

Bardwell, J. L., L. L. Glasgow, and E. A. Epps, Jr. 1962. Nutritional analyses of foods eaten by pintail and teal in south Louisiana. *Proceedings of the Southeastern Association of Game and Fish Commissioners* 16: 209–217.

Burris, F. A. 1991. Diet and behavior of subarctic northern pintails in relation to nutritional requirements of breeding. MS thesis, University of Alaska–Fairbanks, Fairbanks.

Calverley, B. K., and D. A. Boag. 1977. Reproductive potential in parkland and arctic-nesting populations of mallards and pintails (Anatidae). *Canadian Journal of Zoology* 55: 1242–1251.

Cox, R. R. Jr. 1993. Postbreeding ecology of adult male northern pintail and cinnamon teal near Great Salt Lake, Utah. MS thesis, Utah State University, Logan.

Cox, R. R., Jr., and A. D. Afton. 1996. Evening flights of female northern pintails from a major roost site. *Condor* 98: 810–819.

Cox, R. R., Jr., and A. D. Afton. 1997. Use of habitats by female northern pintails wintering in southwestern Louisiana. *Journal of Wildlife Management* 61: 435–443.

Cox, R. R., Jr., A. D. Afton, and R. M. Pace, III. 1998. Survival of female northern pintails wintering in southwestern Louisiana. *Journal of Wildlife Management* 62: 1512–1521.

Derrickson, S. R. 1977. Aspects of breeding behavior in the pintail (*Anas acuta*). PhD dissertation, University of Minnesota–St. Paul, St. Paul.

Derrickson, S. R. 1978. The mobility of breeding pintails. *Auk* 95: 104–114.

DuBowy, P. J. 1980. Optimal foraging and adaptive strategies of post-breeding male blue-winged teal and northern shovelers. MS thesis, University of North Dakota, Grand Forks.

Duncan, D. C. 1987a. Nest-site distribution and overland brood movements of northern pintails in Alberta. *Journal of Wildlife Management* 51: 716–723.

Duncan, D. C. 1987b. Nesting of northern pintails in Alberta: Laying date, clutch size, and renesting. *Canadian Journal of Zoology* 65: 234–246.

Duncan, D. C. 1987c. Variation and heritability in egg size of the northern pintail. *Canadian Journal of Zoology* 65: 992–996.

Esler, D., and J. B. Grand. 1994a. Comparison of age determination techniques for female northern pintails and American wigeon in spring. *Wildlife Society Bulletin* 22: 260–264.

Euliss, N. H., Jr., and S. W. Harris. 1987. Feeding ecology of northern pintails and green-winged teal wintering in California. *Journal of Wildlife Management* 51: 724–732.

Fleskes, J. P., J. L. Yee, G. S. Yarris, M. R. Miller, and M. L. Casazza. 2007. Pintail and mallard survival in California relative to habitat, abundance, and hunting. *Journal of Wildlife Management* 71: 2238–2248.

Fleskes, J. P., R. L. Jarvis, and D. S. Gilmer. 2002. Distribution and movements of female northern pintails radiotagged in San Joaquin Valley, California. *Journal of Wildlife Management* 66: 138–152.

Flint, P. L., and J. B. Grand. 1996a. Nesting success of northern pintails on the coastal Yukon-Kuskokwim Delta, Alaska. *Condor* 98: 54–60.

Flint, P. L., and J. B. Grand. 1996b. Variation in egg size of the northern pintail. *Condor* 98: 162–165.

Flint, P. L., K. Ozaki, J. M. Pearce, B. Guzetti, H. Higuchi, J. P. Fleskes, T. Shimada, and D. V. Derksen. 2009. Breeding-season sympatry facilitates genetic exchange among allopatric wintering populations of northern pintails in Japan and California. *Condor* 111: 591–598.

Fuller, R. W. 1953. Studies in the life history and ecology of the American pintail, *Anas acuta tzitzihoa* (Vieillot), in Utah. MS thesis, Utah State Agricultural College, Logan.

Grand, J. B., and P. L. Flint. 1996. Renesting ecology of northern pintails on the Yukon-Kuskokwim delta, Alaska. *Condor* 98: 820–824.

Guinn, S. J., and B. D. J. Batt. 1985. Activity budgets of northern pintail hens: Influence of brood size, brood age, and date. *Canadian Journal of Zoology* 63: 2114–2120.

Guyn, K. L., and R. G. Clark. 1999. Factors affecting survival of northern pintail ducklings in Alberta. *Condor* 101: 369–377.

Guyn, K. L., and R. G. Clark. 2000. Nesting effort of northern pintails in Alberta. *Condor* 102: 619–628.

Haukos, D. A., M. R. Miller, D. L. Orthmeyer, J. Y. Takekawa, J. P. Fleskes, M. L. Casazza, W. R. Perry, and J. A. Moon. 2006. Spring migration of northern pintails from Texas and New Mexico, USA. *Waterbirds* 29: 127–136.

Hestbeck, J. B. 1993a. Overwinter distribution of northern pintail populations in North America. *Journal of Wildlife Management* 57: 582–589.

Hestbeck, J. B. 1993b. Survival of northern pintails banded during winter in North America, 1950–1988. *Journal of Wildlife Management* 57: 590–597.

Krapu, G. L. 1974a. Feeding ecology of pintail hens during reproduction. *Auk* 91: 278–290.

Krapu, G. L. 1974b. Foods of breeding pintails in North Dakota. *Journal of Wildlife Management* 38: 408–417.

Krapu, G. L., and G. A. Swanson. 1978. Foods of juvenile, brood hen, and post-breeding pintails in North Dakota. *Condor* 79: 504–507.

Krapu, G. L., G. A. Sargeant, and A. E. H. Perkins. 2002. Does increasing daylength control seasonal changes in clutch sizes of northern pintails (*Anas acuta*)? *Auk* 119: 498–506.

Maher, W. J., and D. N. Nettleship. 1968. The pintail (*Anas acuta*) breeding at latitude 82°N on Ellesmere Island, N.W.T., Canada. *Auk* 85: 320–321.

Malecki, R. [A.], S. Sheaffer, D. Howell, and T. Strange. 2006. *Northern Pintails in Eastern North America: Their Seasonal Distribution, Movement Patterns, and Habitat Affiliations: Final Report*. Atlantic Flyway Council Technical Section, US Fish and Wildlife Service.

Mann, F. E., and J. S. Sedinger. 1993. Nutrient-reserve dynamics and control of clutch size in northern pintails breeding in Alaska. *Auk* 110: 264–278.

Migoya, R., and G. A. Baldassarre. 1995. Winter survival of female northern pintails in Sinaloa, México. *Journal of Wildlife Management* 59: 16–22.

Migoya, R., G. A. Baldassarre, and M. P. Losito. 1994. Diurnal activity budgets and habitat functions of northern pintail *Anas acuta* wintering in Sinaloa, México. *Wildfowl* 45: 134–146.

Miller, K. J. 1976. Activity patterns, vocalizations, and site selection in nesting blue-winged teal. *Wildfowl* 27: 33–43.

Miller, M. R. 1985. Time budgets of northern pintails wintering in the Sacramento Valley, California. *Wildfowl* 36: 53–64.

Miller, M. R. 1986a. Northern pintail body condition during wet and dry winters in the Sacramento Valley, California. *Journal of Wildlife Management* 50: 189–198.

Miller, M. R. 1986b. Molt chronology of northern pintails in California. *Journal of Wildlife Management* 50: 57–64.

Miller, M. R. 1987. Fall and winter foods of northern pintails in the Sacramento Valley, California. *Journal of Wildlife Management* 51: 405–414.

Miller, M. R., and D. C. Duncan. 1999. The northern pintail in North America: Status and conservation needs of a struggling population. *Wildlife Society Bulletin* 27: 788–800.

Miller, M. R., J. P. Fleskes, D. L. Orthmeyer, and D. S. Gilmer. 1992. Survival and other observations of adult female northern pintails molting in California. *Journal of Field Ornithology* 63: 138–144.

Miller, M. R., J. P. Fleskes, D. L. Orthmeyer, W. E. Newton, and D. S. Gilmer. 1995. Survival of adult female northern pintails in Sacramento Valley, California. *Journal of Wildlife Management* 59: 478–486.

Miller, M. R., and W. E. Newton. 1999. Population energetics of northern pintails wintering in the Sacramento Valley, California. *Journal of Wildlife Management* 63: 1222–1238.

Miller, M. R., J. Y. Takekawa, J. P. Fleskes, D. L. Orthmeyer, M. L. Casazza, and W. M. Perry. 2005a. Spring migration of northern pintails from California's Central Valley wintering area tracked with satellite telemetry: Routes, timing, and destinations. *Canadian Journal of Zoology* 83: 1314–1332.

Miller, M. R., J. Y. Takekawa, J. P. Fleskes, D. L. Orthmeyer, M. L. Casazza, D. A. Haukos, and W. M. Perry. 2005b. Flight speeds of northern pintails during migration determined using satellite telemetry. *Wilson Bulletin* 117: 364–374.

Nicolai, C. A., P. L. Flint, and M. L. Wege. 2005. Annual survival and site fidelity of northern pintails banded on the Yukon-Kuskokwim Delta, Alaska. *Journal of Wildlife Management* 69: 1202–1210.

Pearse, A. T., G. L. Krapu, R. R. Cox, Jr., and B. E. Davis. 2011. Spring-migration ecology of northern pintails in south-central Nebraska. *Waterbirds* 34: 10–18.

Raveling, D. G., and M. E. Heitmeyer. 1989. Relationships of population size and recruitment of pintails to habitat conditions and harvest. *Journal of Wildlife Management* 53: 1088–1103.

Rice, M. B., D. A. Haukos, J. A. Dubovsky, and M. C. Runge. 2010. Continental survival and recovery rates of northern pintails using band-recovery data. *Journal of Wildlife Management* 74: 778–787.

Richkus, K. D., F. C. Rohwer, and M. J. Chamberlain. 2005. Survival and cause specific mortality of female northern pintails in southern Saskatchewan. *Journal of Wildlife Management* 69: 574–581.

Rienecker, W. C. 1987a. Migration and distribution of northern pintails banded in California. *California Fish and Game* 73: 139–155.

Rienecker, W. C. 1987b. Survival and recovery rate estimates of northern pintails banded in California, 1948–1979. *California Fish and Game* 73: 230–237.

Smith, L. M., and D. G. Sheeley. 1993a. Factors affecting condition of northern pintails wintering in the southern High Plains. *Journal of Wildlife Management* 57: 62–71.

Smith, L. M., and D. G. Sheeley. 1993b. Molt patterns of wintering northern pintails in the southern High Plains. *Journal of Wildlife Management* 57: 229–238.

Smith, R. I. 1968. The social aspects of reproductive behavior in the pintail. *Auk* 85: 381–396.

Smith, R. I. 1970. Response of pintail breeding populations to drought. *Journal of Wildlife Management.* 34: 943–946.

Sorenson, L. G., and S. R. Derrickson. 1994. Sexual selection in the northern pintail (*Anas acuta*): The importance of female choice versus male-male competition in the evolution of sexually selected traits. *Behavioral Ecology and Sociobiology* 35: 389–400.

Sterling, R. T. 1966. Dispersal and mortality of adult drake pintails (*Anas acuta*). MS thesis, University of Saskatchewan, Saskatoon.

Tamisier, A. 1976. Diurnal activities of green-winged teal and pintail wintering in Louisiana. *Wildfowl* 27: 19–32.

Thompson, J. D., and G. A. Baldassarre. 1990. Carcass composition of nonbreeding blue-winged teal and northern pintails in Yucatán, México. *Condor* 92: 1057–1065.

Garganey

Girard, O. 2005. Garganey. Pp. 601–605 in J. Kear, ed. *Ducks, Geese, and Swans.* Vol. 2. Oxford University Press, Oxford, UK.

Owen, M. 1977. Garganey. Pp. 180–183 in *Wildfowl of Europe*, Macmillan, London, UK (Baikal teal, pp. 180–183).

Roselaar, C. S. 1977. Garganey. Pp. 529–536 in S. Cramp and K. E. L. Simmons. *Handbook of the Birds of Europe, the Middle East, and North Africa: The Birds of the Western Palearctic, Volume 1, Ostrich to Ducks.* Oxford University Press, Oxford, UK.

Sugden, L. G. 1963. A garganey duck in the wild in Alberta. *Blue Jay* 21: 4–5.

Blue-winged Teal

Bennett, L. J. 1938. *The Blue-winged Teal: Its Ecology and Management.* Collegiate Press, Ames, IA.

Botero, J. E., and D. H. Rusch. 1994. Foods of blue-winged teal in two Neotropical wetlands. *Journal of Wildlife Management* 58: 561–565.

Burgess, H. H., H. H. Prince, and D. L. Trauger. 1965. Blue-winged teal nesting success as related to land use. *Journal of Wildlife Management* 29: 89–95.

Connelly, J. W., Jr. 1977. A comparative study of blue-winged and cinnamon teal in eastern Washington. MS thesis, Washington State University, Pullman.

Connelly, J. W., and I. J. Ball. 1984. Comparisons of aspects of breeding blue-winged and cinnamon teal in eastern Washington. *Wilson Bulletin* 96: 626–633.

DuBowy, P. J. 1980. Optimal foraging and adaptive strategies of post-breeding male blue-winged teal and northern shovelers. MS thesis, University of North Dakota, Grand Forks.

DuBowy, P. J. 1985a. Feeding ecology and behavior of postbreeding male blue-winged teal and northern shovelers. *Canadian Journal of Zoology* 63: 1292–1297.

DuBowy, P. J. 1985b. Moults and plumages and testicular regression of post-breeding male blue-winged teal (*Anas discors*) and northern shovelers (*Anas clypeata*). *Journal of Zoology* 207: 459–466.

DuBowy, P. J. 1985c. Seasonal organ dynamics in post-breeding male blue-winged teal and northern shovelers. *Comparative Biochemistry and Physiology* 82A: 899–906.

Gammonley, J. H., and L. H. Fredrickson. 1995. *Life History and Management of the Blue-winged Teal.* Waterfowl Management Handbook 13.1.8. National Biological Service, Washington, DC.

Geis, A. D., R. I. Smith, and S. V. Goddard. 1963. *Blue-winged Teal Band Recovery and Annual Mortality Rates.* Administrative Report 18. Migratory Bird Populations Station, Branch of Wildlife Research, US Bureau of Sport Fisheries and Wildlife, Laurel, MD.

Glover, F. A. 1956. Nesting and production of the blue-winged teal (*Anas discors* Linnaeus) in northwest Iowa. *Journal of Wildlife Management* 20: 28–46.

Harris, H. J., Jr. 1970. Evidence of stress response in breeding blue-winged teal. *Journal of Wildlife Management* 34: 747–755.

Lewis, T. E., and P. R. Garrettson. 2010. Parasitism of a blue-winged teal nest by a northern shoveler in South Dakota. *Wilson Journal of Ornithology* 122: 612–614.

Loos, E. R. 1999. Incubation in blue-winged teal (*Anas discors*): Testing hypotheses of incubation constancy, recess frequency, weight loss, and nest success. MS thesis, Louisiana State University, Baton Rouge.

McHenry, M. G. 1971. Breeding and post-breeding movements of blue-winged teal (*Anas discors*) in southwestern Manitoba. PhD dissertation, University of Oklahoma, Norman.

Miller, K. J. 1976. Activity patterns, vocalizations, and site selection in nesting blue-winged teal. *Wildfowl* 27: 33–43.

Mulhern, J. H., T. D. Nudds, and B. R. Neal. 1985. Wetland selection by mallards and blue-winged teal. *Wilson Bulletin* 97: 473–485.

Ringelman, J. K., and L. D. Flake. 1980. Diurnal visibility and activity of blue-winged teal and mallard broods. *Journal of Wildlife Management* 44: 822–829.

Rollo, J. D., and E. G. Bolen. 1969. Ecological relationships of blue and green-winged teal on the High Plains of Texas in early fall. *Southwestern Naturalist* 14: 171–188.

Sharp, B. 1972. Eastward migration of blue-winged teal. *Journal of Wildlife Management* 36: 1273–1277.

Stewart, R. E., and J. W. Aldrich. 1956. Distinction of maritime and prairie populations of blue-winged teal. *Proceedings of the Biological Society of Washington* 69: 29–34.

Strohmeyer, D. L. 1967. The biology of renesting by the blue-winged teal (*Anas discors*) in northwest Iowa. PhD dissertation, University of Minnesota, Minneapolis.

Swanson, G. A., and M. I. Meyer. 1977. Impact of fluctuating water levels on feeding ecology of breeding blue-winged teal. *Journal of Wildlife Management* 41: 426–433.

Swanson, G. A., M. I. Meyer, and J. R. Serie. 1974. Feeding ecology of breeding blue-winged teals. *Journal of Wildlife Management* 38: 396–407.

Taylor, T. S. 1978. Spring foods of migrating blue-winged teal on seasonally flooded impoundments. *Journal of Wildlife Management* 42: 900–903.

Thompson, J. D., and G. A. Baldassarre. 1990. Carcass composition of nonbreeding blue-winged teal and northern pintails in Yucatán, México. *Condor* 92: 1057–1065.

Wheeler, R. J. 1965. Pioneering of the blue-winged teal in California, Oregon, Washington, and British Columbia. *Murrelet* 46: 40–42.

White, D. H., K. A. King, C. A. Mitchell, and A. J. Krynitsky. 1981. Body lipids and pesticide burdens of migrant blue-winged teal. *Journal of Field Ornithology* 52: 23–28.

Wilson, R. E. 2008. Genetic and phenotypic divergence within and between cinnamon and blue-winged teal. PhD dissertation, University of Alaska, Fairbanks.

Wilson, R. E., M. Eaton, S. A. Sonsthagen, J. L. Peters, K. P. Johnson, B. Simarra, and K. G. McCracken. 2011. Speciation, subspecies divergence, and paraphyly in the cinnamon teal and blue-winged teal. *Condor* 113: 747–761.

Cinnamon Teal

Bolen, E. G. 1978. Notes on blue-winged teal × cinnamon teal hybrids. *Southwestern Naturalist* 23: 692–696.

Connelly, J. W., Jr. 1977. A comparative study of blue-winged and cinnamon teal in eastern Washington. MS thesis, Washington State University, Pullman.

Connelly, J. W., Jr. 1978. Trends in blue-winged teal and cinnamon teal populations in eastern Washington. *Murrelet* 59: 2–6.

Connelly, J. W., and I. J. Ball. 1984. Comparisons of aspects of breeding blue-winged and cinnamon teal in eastern Washington. *Wilson Bulletin* 96: 626–633.

Cox, R. R., Jr. 1993. Postbreeding ecology of adult male northern pintails and cinnamon teal near Great Salt Lake, Utah. MS thesis, Utah State University, Logan.

Gammonley, J. H. 1995a. Nutrient reserve and organ dynamics of breeding cinnamon teal. *Condor* 97: 985–992.

Gammonley, J. H. 1995b. Spring feeding ecology of cinnamon teal in Arizona. *Wilson Bulletin* 107: 64–72.

Hohman, W. L. 1991. Incubation rhythm components for three cinnamon teal nesting in California. *Prairie Naturalist* 23: 229–233.

Hohman, W. L., and C. D. Ankney. 1994. Body size and condition, age, plumage quality, and foods of prenesting cinnamon teal in relation to pair status. *Canadian Journal of Zoology* 72: 2172–2176.

Kozlik, F. M. 1972. California Pacific Flyway report, second and third quarters, 1972. *Pacific Flyway Report 68*, Sacramento, CA.

Lokemoen, J. T., and D. E. Sharp. 1981. First documented cinnamon teal nesting in North Dakota produced hybrids. *Wilson Bulletin* 93: 403–405.

McKinney, F. 1970. Displays of four species of blue-winged ducks. *Living Bird* 9: 29–64.

Snyder, L. L., and H. G. Lumsden. 1951. Variation in *Anas cyanoptera*. Royal Ontario Museum of Zoology, *Occasional Papers* 10: 1–18.

Spencer, H. E., Jr. 1953. The cinnamon teal (*Anas cyanoptera* Vieillot): Its life history, ecology, and management. MS thesis, Utah State University, Logan.

Stark, R. S. 1979. Morphological differences between blue-winged and cinnamon teal. MS thesis, Colorado State University, Fort Collins.

Thorn, T. D., and P. J. Zwank. 1993. Foods of migrating cinnamon teal in central New Mexico. *Journal of Field Ornithology* 64: 452–463.

Wilson, R. E., M. Eaton, and K. G. McCracken. 2008. Color divergence among cinnamon teal (*Anas cyanoptera*) subspecies from North America and South America. *Ornitología Neotropical* 19: 307–314.

Wilson, R. E., M. Eaton, and K. G. McCracken. 2012. Plumage and body size differentiation in blue-winged teal and cinnamon teal. *Avian Biology Research* 5: 107–116.

Wilson, R. E., T. H. Valqui, and K. G. McCracken. 2010. Ecogeographic variation in cinnamon teal (*Anas cyanoptera*) along elevational and latitudinal gradients. *Ornithological Monographs* 67: 141–161.

Wallace, K. L. M., and M. A. Ogilvie. 1977. Distinguishing blue-winged and cinnamon teals. *British Birds* 70: 290–294.

Weseloh, D. V., and M. Weseloh. 1979. Probable hybrids of cinnamon × blue-winged teal from southern Alberta. *Canadian Field-Naturalist* 93: 316–317.

Wilson, R. E. 2008. Genetic and phenotypic divergence within and between cinnamon and blue-winged teal. PhD dissertation, University of Alaska, Fairbanks.

Wilson, R. E., M. Eaton, S. A. Sonsthagen, J. L. Peters, K. P. Johnson, B. Simarra, and K. G. McCracken. 2011. Speciation, subspecies divergence, and paraphyly in the cinnamon teal and blue-winged teal. *Condor* 113: 747–761.

Wilson, R. E., T. H. Valqui, and K. G. McCracken. 2010. Ecogeographic variation in cinnamon teal (*Anas cyanoptera*) along elevational and latitudinal gradients. *Ornithological Monographs* 67: 141–161.

Northern Shoveler

Afton, A. D. 1977. Aspects of reproductive behavior in the northern shoveler. MS thesis, University of Minnesota, St. Paul.

Afton, A. D. 1979a. Time budget of breeding northern shovelers. *Wilson Bulletin* 91: 42–49.

Afton, A. D. 1979b. Incubation temperatures of the northern shoveler. *Canadian Journal of Zoology* 57: 1052–1056.

Afton, A. D. 1980. Factors affecting incubation rhythms of northern shovelers. *Condor* 82: 132–137.

Ankney, C. D., and A. D. Afton. 1988. Bioenergetics of breeding northern shovelers: diet, nutrient reserves, clutch size, and incubation. *Condor* 90: 459–472.

Boyer, G. F. 1949. Breeding of the shoveler in New Brunswick. *Auk* 66: 199–200.

DuBowy, P. J. 1980. Optimal foraging and adaptive strategies of post-breeding male blue-winged teal and northern shovelers. MS thesis, University of North Dakota, Grand Forks.

DuBowy, P. J. 1985a. Feeding ecology and behavior of postbreeding male blue-winged teal and northern shovelers. *Canadian Journal of Zoology* 63: 1292–1297.

DuBowy, P. J. 1985b. Moults and plumages and testicular regression of post-breeding male blue-winged teal (*Anas discors*) and northern shovelers (*Anas clypeata*). *Journal of Zoology* 207: 459–466.

DuBowy, P. J. 1985c. Seasonal organ dynamics in post-breeding male blue-winged teal and northern shovelers. *Comparative Biochemistry and Physiology* 82A: 899–906.

DuBowy, P. J. 1997. Long-term foraging optimization in northern shovelers. *Ecological Modelling* 95: 119–132.

Fuller, R. W., and N. E. King. 1964. American wigeon and shoveler breeding in Vermont. *Auk* 81: 86–87.

Griffith, R. E. 1946. Nesting of gadwall and shoveler on the middle Atlantic coast. *Auk* 63: 436–438.

Guillemain, M., H. Fritz, and N. Guillon. 2000. Foraging behavior and habitat choice of wintering northern shoveler in a major wintering quarter in France. *Waterbirds* 23: 355–363.

Hori, J. 1962. The pre-nuptial display of the shoveler. *Wildfowl* 13: 173–174.

Lewis, T. E., and P. R. Garrettson. 2010. Parasitism of a blue-winged teal nest by a northern shoveler in South Dakota. *Wilson Journal of Ornithology* 122: 612–614.

MacCluskie, M. C., and J. S. Sedinger. 1999. Incubation behavior of northern shovelers in the subarctic: a contrast to the prairies. *Condor* 101: 417–421.

MacCluskie, M. C., and J. S. Sedinger. 2000. Nutrient reserves and clutch-size regulation of northern shovelers in Alaska. *Auk* 117: 971–979.

McKinney, F. 1967. Breeding behaviour of captive shovelers. *Wildfowl Trust Annual Report* 18: 108–121.

Payn, W. H. 1941. The plumage changes of adolescent shovelers. *Ibis* 83: 456–459.

Poston, H. J. 1969. Relationships between the shoveler and its breeding habitat at Strathmore, Alberta. Pp. 132–137 in *Saskatoon Wetlands Seminar: Transactions of a Seminar on Small Water Areas in the Prairie Pothole Region*. Report Series 6. Department of Indian Affairs and Northern Development, Canadian Wildlife Service, Ottawa, ON.

Poston, H. J. 1974. *Home Range and Breeding Biology of the Shoveler*. Report Series 25. Canadian Wildlife Service, Ottawa, ON.

Ross, R. K., and N. R. North. 1983. Breeding records of northern shovelers, *Anas clypeata*, along the northern coast of Ontario. *Canadian Field-Naturalist* 97: 113.

Seymour, N. R. 1974a. Territorial behaviour of wild shovelers at Delta, Manitoba. *Wildfowl* 25: 49–55.

Seymour, N. R. 1974b. Site attachment in the northern shoveler. *Auk* 91: 423–427.

Seymour, N. R. 1974c. Aerial pursuit flights in the shoveler. *Canadian Journal of Zoology* 52: 1473–1480.

Stewart, P. A. 1957. Nesting of the shoveller (*Spatula clypeata*) in central Ohio. *Wilson Bulletin* 69: 280.

Tietje, W. D. 1986. Aspects of the wintering ecology of northern shovelers on freshwater and saline habitats. PhD dissertation, Texas A&M University, College Station.

Tietje, W. D., and J. G. Teer. 1988. Winter body condition of northern shovelers on freshwater and saline habitats. Pp. 353–376 in M. W. Weller, ed. *Waterfowl in Winter*. University of Minnesota Press, Minneapolis.

www.ingramcontent.com/pod-product-compliance
Lightning Source LLC
Chambersburg PA
CBHW080610270326

41928CB00016B/2987